INSTITUTIONS, POLICY AND OUTPUTS FOR ACIDIFICATION

# Institutions, Policy and Outputs for Acidification

The Case of Hungary

LAURENCE J. O'TOOLE, Jr
*University of Georgia*
*USA*

*Studies in Green Research*

Aldershot • Brookfield USA • Singapore • Sydney

Published by
Ashgate Publishing Ltd
Gower House
Croft Road
Aldershot
Hants GU11 3HR
England

Ashgate Publishing Company
Old Post Road
Brookfield
Vermont 05036
USA

**British Library Cataloguing in Publication Data**

O'Toole, Laurence J., 1948-
Institutions, policy and outputs for acidification : the case of Hungary
1.Acid rain - Hungary 2.Acidification
I. Title
363.7'386'09439

**Library of Congress Catalog Card Number:** 97-76939

ISBN 1 85972 611 9

Printed and bound by
Antony Rowe Ltd, Chippenham, Wiltshire

# Contents

# Figures and tables

# Preface

This volume reports the results of a study of the implementation of one set of international environmental agreements in one nation over a period of approximately ten years. The agreements are those stemming from the Convention on Long Range Trans-boundary Air Pollution (CLRTAP) of the Economic Commission of Europe and encompass a set of protocols adopted by many countries. The emphasis on implementa-tion, as opposed to a concentration on adoption, reflects the choice to focus on the following research question: do such agreements make a demonstrable impact 'on the ground' - or in this case, in the air? The nation receiving attention here is Hungary, both before and during the political and economic transformations that have been so consequential in recent years.

This subject is best approached by building upon and integrating multiple perspectives. The analysis in this book draws from the ideas and insights of three streams of scholarship: the systematic study of policy implementation; the emphasis in international relations on understanding the role and significance of international regimes like those addressing environmental issues; and the analysis of the so-called 'countries in transition', the nations undergoing shifts from state-socialist systems to democratic and market-based arrangements in Central and Eastern Europe. Each of these subjects is a specialty on its own, and each has developed largely in isolation from the others. Combining them, nonetheless, seems an appropriate strategy for understanding the impact of CLRTAP in Hungary and, in complementary fashion, enhancing theory building in these fields.

A good deal of my recent research has been devoted to the analysis of programs aimed at protecting and improving the environment, a subject of intense and growing concern in many countries. As a long-time student of policy implementation, as well, I have come to be aware of the strengths and also the limitations of some conventional frameworks and theories of that specialty. I have become convinced that, particularly for policy problems like those crossing governments' jurisdictional lines, so-called 'top-down' perspectives offer at most highly incomplete analytical approaches. Indeed, these formulations can be positively misleading; they suggest that the solution to policy challenges can be found in strong authority, tight control, close monitoring, and substantial sanctions. And yet - other objections aside - for international environmental policy issues there is no strong common authority across the relevant parties, tight control is likely to be a mirage, close monitoring is no mean feat, and the application of serious sanctions is often infeasible. Top-down perspectives, in short, provide little guidance either in explaining the range of variation across extant cases or in assisting actors in the world of practice in improving the utility, implementability, and effectiveness of international agreements. The implementation of international environmental agreements is thus attractive as a subject through which ideas about implementation beyond hierarchy can be developed and refined.

An additional concern with the limitations of recent theoretical ideas about policy implementation stimulated my interest in pursuing this kind of investigation in Central and Eastern Europe in particular. A few years ago, I explored issues of implementation in another substantive area, privatization policy, in the early years of Hungary's post-socialist transition. (As it turns out, the privatization effort is connected to parts of the implementation experience for air quality - as the present book makes clear.) That work convinced me that frameworks and theories of policy implementation, as developed in Western Europe and the United States, carry implicit premises - regarding regime stability and widespread commitment to the principles undergirding liberal systems - that do not necessarily hold elsewhere. In particular, I found that standard perspectives on processes of implementation do not explain much that is of interest in transitional settings like that in Hungary, and it can be important to think afresh about processes of coordinating and concerting action. Here, after all, public programs are adopted by and through national regimes that are in substantial flux.

This finding, nevertheless, suggests some cautionary points worth emphasizing at the outset. This volume consists of the study of a case -

or, for comparisons across protocols developed under CLRTAP, a set of closely related cases. The detailed examination over a period of years clearly offers advantages. But generalizing from this experience to the universe of international environmental agreements is hazardous. Furthermore, Hungary is one nation, and hardly a typical one at that. While a study of the Hungarian experience can be highly revealing about both the limitations of conventional perspectives and also the subject of implementation under transitional conditions, the results of this research must be treated with circumspection by those interested in other countries and other conditions. Indeed, as explained more carefully in the body of this volume, the unusual nature of recent Hungarian experience - with many change processes developing simultaneously - creates difficulties in determining which elements of the international agreement and regime, or even which domestic influences, may be responsible for any observable results - in terms of air quality or compliance with international commitments, for example. Still, most scholarship on the role of such agreements has concentrated on Western nations and the development of national commitments - as opposed to the impacts of formal agreements on patterns and practices *within* participating nations. This book extends this line of research to the important Central and Eastern European region and into the crucial 'details' of implementation. It may be of interest to specialists on several topics, including environmental policy and politics, the politics and economics of the Central and Eastern Europe, international organizations and international environmental agreements, policy implementation, and air quality and acidification.

Almost any book is in some sense a collective endeavor. This one certainly is. The research on which this volume is based was supported by the European Commission through two projects: a Social and Economic aspects of Environmental Research, or SEER, effort on 'International Organizations and National Participation in Environmental Regimes: The Organizational Component of the Acidification Regime', under EC Contract EV5V-CT94-0390; and the related initiative on 'International Organizations and National Participation in Environmental Regimes: Hungarian Involvement in the Acidification Regime', supported by the program on Cooperation in Science and Technology with Central and Eastern European Countries, Directorate-General XII, Science, Research and Development (PECO Contract number ERBCIPDCT94109). Without this assistance during 1994-96, this work could never have been conducted.

The Commission's PECO program aims to encourage cooperative efforts between those in the West and colleagues in Central and Eastern Europe, and this project benefitted from just such collaboration. The report submitted to the Commission, a product titled 'Hungary: Institutions, Policy, and Outputs for Acidification' and which has been revised to its present form for book publication, was written by me. But the results are based on support and collaboration from colleagues in Hungary. Zsuzsanna Flachner, now of the Ministry for Environment and Regional Policy of the Government of Hungary, assisted in research and contributed materially by finding and interpreting some of the raw materials for analysis, translating and analyzing during the course of the investigation, and providing research advice and suggestions for improving the manuscript. Professor Sándor Kerekes of the Budapest University of Economic Sciences offered additional help by arranging some interviews, offering generous use of documents, and providing administrative oversight for the work in Hungary.

Kenneth Hanf of the Erasmus University Rotterdam, The Netherlands, was the overall coordinator for the SEER and companion SEER/PECO projects. He provided leadership for this initiative as well as for related research efforts underway in several other countries. Indeed he had instigated an earlier successful SEER study on 'The Domestic Basis of International Environmental Agreements: Modelling National/International Linkages' (see report to the European Commission for EC Contract EVSV-CT92-0185, May 1996) which provided valuable background for the current investigation. Ken performed many important functions with admirable skill: he managed the project and served as liaison with the EC, helped identify appropriate Hungarian colleagues for collaboration, offered ideas and suggested revisions, and kept the rest of us smiling and productive through the period of research.

Thanks are due as well to Monika Andrási in Budapest, who provided assistance with some data gathering and translations during the earliest phase of this work; and both Stephanie Steen and Richard Schaeffer, graduate students at the University of Georgia, who helped with literature research. Richard also assisted with numerous details of manuscript production.

A considerable portion of the information needed for the analysis offered in this volume derives from interviews with and other data provided by experts and officials in Hungary. These sources are listed at the conclusion of the study, and their cooperation - in some cases, extending to several discussions over a considerable period - has been invaluable in developing the results of this inquiry. Numerous officials and practitioners in Hungary also read and offered helpful comments on an earlier draft of this book. Several attended a full-day discussion of the draft and its implications, at a workshop convened at the Budapest University of Economic Sciences on 2 April 1996. The meeting was supported by the PECO Program of the European Union. Thanks are due to these individuals and this Program.

Some of the data used in the development of the analysis are drawn from interviews that were assisted with instruments developed by Detlef Sprinz and Carsten Helm of the Potsdam Institute for Climate Change Research in Germany, who helped to organize a complementary several-nation multivariate analysis. Their work as well has been supported by the SEER project on International Organizations and National Participation in Environmental Regimes.

Arild Underdal of the University of Oslo read an earlier version of this work and offered helpful comments.

A portion of the coverage of institutional capacity of Hungarian institutions for environmental policy implementation is reprinted here by permission of the journal *Environmental Politics* and its publisher, Frank Cass. The material is drawn from an article co-authored with Kenneth Hanf and forthcoming in a special issue on environmental issues in Central and Eastern Europe. I acknowledge with gratitude the publisher's permission to reprint.

Finally, my family deserves heartfelt thanks. Mary, Conor, and Katie tolerated absences, preoccupations, frayed nerves, false starts, as well as bouts of euphoria - all with calm support and exceptionally good humor - during research, writing, and revision. To this sustaining trio I owe the most.

Athens, Georgia
July 1997

# 1 Introduction

Roughly 175 international agreements addressing environmental issues currently operate throughout the globe. Challenging as it often is to induce nations' approval of the terms of such agreements, the task of converting formal support into successfully implemented streams of action can involve a particularly vexing and difficult set of tasks.

Scholars and practitioners have paid increasing attention to the development of these international agreements (see, for instance, Kimball 1992; Sand 1990; Susskind 1994; Young 1989). And analytic efforts have started to emerge with the aims of understanding the complications of implementation and developing lessons useful for improving performance (Brown and Churchill 1985; Bryner 1996; Chayes and Chayes 1993; Hanf and Underdal 1996; Jacobson and Weiss 1997; Keohane, Haas, and Levy, 1993; Victor, Raustiala, and Skolnikoff, forthcoming; and Young and von Moltke 1994). A wide variety of research approaches have been marshal-led in this cause. It is clear that, while some lines of convergence and some accumulated wisdom have begun to develop, considerably more work is necessary to understand the full range of influences on performance. In this regard, empirical research is extremely important, with particular attention to the substantial and growing set of experiences with the operating agreements - investigations on multiple environmental issues, in multiple locales, and over time.

This book offers one study to contribute to this effort. The approach is a one-country focussed case investigation of more than a decade's experience with agreements regarding one international environmental challenge. The volume reports on an analysis of institutions, policy and policy implementation, and outputs for dealing with acidification[1] in Hungary, as well as links between domestic Hungarian developments in this field and the relevant international regimes - most especially the Convention on Long Range Trans-boundary Air Pollution (CLRTAP or LRTAP) of the Economic Commission of Europe. It is based primarily on reviews of relevant documents, literature, unpublished data and files in Hungary, and numerous interviews with officials, analysts, and other stakeholders in Hungary, primarily in Budapest (see the list of Persons Interviewed included at the end of this book). Field research was conducted between October 1994 and March 1996.

**LRTAP and the Hungarian Context**

The Convention on Long Range Trans-boundary Air Pollution stands as one of the earliest and most visible international environmental agreements, particularly in Europe. It was initiated following the Helsinki Conference on Security and Cooperation in Europe of 1975, when the Soviet Union proposed a high-level meeting of East and West representatives to consider subjects for potential cooperation (see Levy 1993). The Economic Commission of Europe (ECE), as one of the regional economic commissions of the United Nations, had the task of identifying a promising subject for the planned international conference. The ECE selected the environment as the proposed topic of collaborative effort and eventually settled on air pollution as the focus of international discussion. A convention was negotiated during 1978-1979 via ECE working groups.

The countries of the East were motivated to participate primarily by interest in the détente agenda, while Nordic countries and some West European states had developed serious concerns about air pollution. Despite some opposition, especially for a time from Germany, agreement on a convention was reached by 1979 among 33 nations, including Canada and the United States - both countries participating as members of the ECE.

LRTAP did not immediately establish regulatory provisions; nor did it delineate specific, costly and mandatory changes in nations' behavior for the future. But the convention did enshrine the principle that transboundary air pollution should be minimized as much as economically feasible, establish some information collection and dissemination procedures, and initiate a modest institutional mechanism for administering the information function and the development of future protocols specifying reductions (Levy 1993, p.83). The secretariat activities of LRTAP are handled by the Air Pollution Unit of the Environment and Human Settlements Division of the ECE, and the Unit devotes almost all its efforts to organizing meetings convened under the convention's framework. Policy making is the jurisdiction of the Executive Body (EB), which consists of officials from the signatory countries. Specific tasks requiring decision are distributed to subsidiary working groups, which draft regulatory protocols and supervise the operations of collaborative research efforts (Levy 1993, pp.84-85). These research groups and various task forces act under the jurisdictions of the working groups and are open to those interested in participating.

> The EB is constituted 'within the framework' of the ECE's Senior Advisers on Environmental Problems, yet is formally independent of the ECE. Because virtually no decisions are made by a secretariat, but rather by national officials representing their governments, LRTAP has been accurately called a 'permanent negotiating process'. [Levy 1993, p.85, quoting Fauteux 1990, p.6]

Ratification by 16 nations, of the 33 involved at the outset, was required to enter LRTAP requirements into force. Several protocols have been enacted and others are currently under negotiation or discussion. Specifics in the most important of these for the period under examination in this volume are presented later. In practice, nations not supportive of one or more protocols refrain from participating in the decision making and subsequent implementation. Representatives of nongovernmental organizations (NGOs), industry, and international organizations are also permitted to participate in LRTAP gatherings.

In the early years of this process, the impetus in favor of controlling acidification came primarily from certain of the countries of Western Europe, including in Scandinavia. But formal support was not limited to that region. Hungary was one of the original countries involved in the establishment of this acidification regime. In a pattern established during the state-socialist period and continued to the present, Hungarian leaders signed and ratified the Convention; approved a subsequent protocol establishing a measuring system for transboundary air pollution; and ultimately supported several emissions protocols, including two for reducing sulfur dioxide, one controlling nitrogen oxides, and another dealing with volatile organic compounds. Hungary has also been an active participant in research efforts and other LRTAP activities and through its representatives has acted at the international level as a nation seriously interested in the acidification regime and its purposes.

The reasons for this support, the extent to which the acidification regime has had real impacts within Hungary, and the ways in which Hungarian experience and preferences have influenced actions at the international level are the main subjects of this book. Indeed, since understanding Hungarian behavior in connection with LRTAP requires knowledge of Hungarian policy and institutions for environmental questions, particularly air pollution, a secondary objective of this volume is to provide a characterization of Hungary on these matters more broadly.

## Organization of the Book

The emphasis of this investigation is on the highly practical issues surrounding internationally-encouraged efforts to control and reduce pollution from the chemical emissions targeted by programs aimed at the problem of acid rain. The study emphasizes the social-scientific aspects of this topic, as applied in the case of Hungary. The core of this effort is to seek understanding of whether and how international agreements for reducing acidification have impacts on behavior, through their influence on a range of social institutions and processes.

In particular, the book focuses on:

- Whether and how the principal international regime in this policy sector has influenced national decision making in Hungary, and how Hungary has approached and sought to influence the evolving international regime;
- Whether and how national policy in Hungary has affected policy implementation on controlling the relevant pollutants and has encouraged supportive behavior among a range of other social actors, from manufacturing companies to local governments; and, in turn,
- How policy and implementation experience may feed into and influence international negotiations.

Therefore, although emission levels are relevant, establishing whether Hungary has complied with international agreements regarding acidification is only one part of the picture. The investigation also explores governmental input into international negotiations, as well as subsequent impacts on Hungarian policy of the accumulated international agreements; the complications involved in trying to convert official Hungarian policy on issues related to acidification into streams of supportive action during implementation; socioeconomic and political repercussions of the implementation efforts, and their ultimate impacts on the world of action; and processes of learning and feedback as the actors involved in the Hungarian experience operate between national and international levels in reformulating the negotiated understandings.

Needless to say, establishing such linkages must rely on inferences made from the data available. Investigating not merely regime impact on emissions levels - a task itself difficult enough under usual circumstances - but also ramifications through processes of policy action requires relying on the perspectives and interpretations of observers and participants, not simply a set of unobtrusive data. These actors inevitably disagree on some important issues; so a residuum of judgment remains, inevitably, with the researcher in an inquiry such as this one.

The conclusions sketched in this volume can be expected to be clearer and more certain regarding impacts of the CLRTAP on social institutions and processes than on the issue of ultimate concern: the chemical emissions and air quality in Hungary and its region. The point may be put differently. As analysts like Bernauer have noted, studies of the effects of international environmental institutions have typically been fairly loose and ambiguous regarding the specification of the dependent variable, and the causal paths by which such institutions influence results have been ignored or inadequately documented (1995). As the title of this volume suggests, the dependent variables of interest here are several, including institutional and policy impacts, along with the outputs of policy action: the products of governmental effort, possibly including collaborative initiatives with other social actors. The net effect of these streams of effort on air quality - the *outcomes* traceable to LRTAP - is a more disputatious subject. Nevertheless, although this issue is not explicitly advertised in this book's title, it too is a focus of some analytical effort here, and conclusions are also offered regarding this central question.

An additional phenomenon renders this study especially challenging. For Hungary during the period under study - roughly from the late 1970s to the present - a set of centrally important social, economic, and political changes complicate any causal inferences that might be made. The methodological problem here is the classic difficulty plaguing one-case longitudinal investigations when multiple plausibly-independent variables are in play. The complications make the Hungarian experience more interesting and important, albeit more difficult to analyze. This issue is treated more carefully below.

The book is organized as follows. Chapter 2 frames the investigation by providing basic information about Hungary and some of the country's current challenges, as these are relevant for the subsequent analysis. That chapter also includes an explanation of the significance of the Hungarian case for those interested more broadly in understanding the influence of international environmental agreements in general, and the acidification regime in particular. For the Hungarian case is unusual, when compared with most of the nations that have more typically been the focus of research in this field.

Chapter 3 also establishes context, in this case by outlining some of the more important hypothesized causal forces that could be at work influencing the success of Hungary's efforts to make progress on acidification issues. Since several social processes may plausibly be expected to influence acidification outputs and outcomes, it is important to identify some of the most important of these. Attention is focussed in particular on hypothesized influences that are relatively unusual, at least by comparison with countries in Western Europe during the same period. Emphasis is given to the recent domestic political changes, property shifts and the introduction of market forces into the Hungarian economy, and pressures for European integration. The hypothesized influences on action regarding acidification are thus framed at the outset and revisited at the conclusion of the volume.

Chapter 4 turns attention to the state of the environment, and air quality in particular, in Hungary. Coverage examines the environment as an issue of domestic concern, including political concern. Recent and current developments regarding air quality and acidification concerns are also highlighted. The next two chapters shift the attention to policy and its enforcement: Chapter 5 for environmental policy generally, Chapter 6 for air pollution policy, which is obviously of central concern for this research.

Because the ability to carry out policy commitments is often a crucial limitation of policy making, and not simply in Hungary, Chapters 7 and 8 explicate the system for organizing and executing government programs and initiatives. The former chapter focuses on questions of organization, administration, and policy implementation, while the latter addresses the complex issue of institutional capacity and capacity building, important themes in Central and Eastern Europe in recent years.

Chapters 9 and 10 then analyze many features of the Hungarian approach to LRTAP. The principal actors, institutions, events, and processes during the years before and after the political changes of 1990 are covered. Emphasis is on accomplishing two tasks: sketching the efforts involved to convert international stimuli regarding the acidification question into practical impacts within Hungary, and also identifying evidence that can help to answer the central question of whether this international regime has made any difference in the Hungarian instance. Chapter 10 includes some information on processes and results for each of the several protocols that have been developed or are developing as part of the acidification regime.

Chapter 11 then examines the implementation efforts and gaps in Hungarian attempts to carry out its acidification decisions. And Chapter 12 reviews the main findings regarding LRTAP and offers answers to the central research questions. This treatment allows for consideration of more aspects of the domestic-international linkage and also includes coverage of the issue of policy-oriented learning, as the matter pertains to the Hungarian case.

Chapter 13 then devotes brief attention to the link between Hungary and another international regime, the European Union, as this subject is relevant to acidification; and longer-term influences within the Hungarian policy community of the multiple and increasing international commitments.

Finally, a Conclusion revisits some of the hypothesized influences raised at the outset, considers the effectiveness of the LRTAP regime in light of broader scholarship on international environmental agreements, and suggests principal challenges to Hungarian efforts for the future.

## Note

1 The terms 'acidification' and 'acid rain' are used here in a broad sense, to include long distance transport of pollutants by air. Some such pollutants, such as sulfur dioxide and nitrogen oxides, are largely responsible for acidification and are believed to constitute the most important ecological threats among the range of pollutants travelling across national borders. These materials were the initial focus of scientific and policy discussion, both within European countries and in international forums. More recent pollutants of interest within the 'acidification regime' are, strictly speaking, threats to environmental quality for reasons apart from chemical acidification effects. These include volatile organic compounds of several types (VOCs), heavy metals, and other airborne substances. The full set of pollutants and the policies and institutional arrangements developed to deal with them are referred to in this report in terms of the simpler

notion of acidification. This usage accords with the standard terminology in the field, even though concerns in Hungary have been focussed primarily on human health, and also on the loss of agricultural productivity and the general matter of air quality, rather than on some of the acidification issues important in the Scandinavian countries and elsewhere.

# 2 The Hungarian setting

In many respects Hungary is a particularly interesting setting in which to explore domestic-international linkages on environmental policy and acidification issues. This chapter provides some basic information about the national context, summarizes some important recent political and economic developments, and places the case in the context of the international regime of particular interest in this investigation: the Convention on Long Range Trans-boundary Air Pollution (CLRTAP) of the Economic Commission of Europe.

## The Setting: Hungary in the Midst of Political Transformation and Environmental Challenge

Hungary, a landlocked country in Central Europe, is a nation of approximately 10.6 million, 20 percent of whom live in Budapest, the capital. The remainder of the population is located in a number of smaller cities, some of which approach perhaps ten percent of Budapest's population; villages; and rural areas.

Hungary is a nation rich in agricultural resources, especially arable land which comprises approximately 88 percent of the country (see Ministry for Environment and Regional Policy, 1991b, p.33); adequately endowed with water supplies, at least at present; fairly poor in natural resources like minerals; and moderately industrialized.

While Hungary has not experienced some of the severely devastating environmental catastrophes that some of its neighbors have had to face after decades of state socialism, there are a number of regions in the country where serious environmental problems have been identified and have been matters for discussion, particularly within national government and in some local communities, as well as occasionally with non-governmental organizations (NGOs).

The relatively small size of the country (93,030 square kilometers) and the geographic/topographic setting mean that Hungary is heavily interdependent in ecological terms with its neighbors, especially the contiguous nations of Austria, Slovakia, the Czech Republic, Croatia, Serbia, Romania, and the Ukraine. The interdependence is consequential for several environmental media and has occasionally resulted in high-visibility international conflicts and negotiations. The most prominent among these has been the Gabcikovo-Nagymaros Dam issue, which involved the post-socialist government (see below) in a decision to withdraw from participation in joint construction of the Danube River-based project with Slovakia (then Czechoslovakia).

For water, in fact, Hungary is particularly reliant on supplies from abroad: surface water derives 94 percent from other countries (Ministry for Environment and Regional Policy, 1991b, p.4). The increasing levels of contamination of ground water make this fact of growing importance. Indeed, threats to water quality are currently seen as the leading environmental challenge - unlike the situation in some neighboring countries, where combatting air pollution constitutes the first priority. The ecological interdependence extends beyond water issues, which have attracted the most domestic and international attention in the case of Hungary, to air pollution and acid rain. And the full set of environmental issues are affected as well by the processes of political and economic transformation currently underway.

## Political Transformation

The political changes in Central and Eastern Europe beginning in 1989, which culminated within Hungary in a new regime and a fledgling democratic system, have drastically altered the context in which environmental policy and acidification efforts are developed and executed.

The former state-socialist system was known to be somewhat more open to the West than were regimes in some of the surrounding countries, but the Hungarian setting for political expression and policy making was different in important respects from systems to the West. Among the features that bore on the issues of concern for the present study were the following: limitations placed on political expression, gradually giving way to environmentally-focussed political protest in the 1980s; one-party central governance, with little overt political conflict and therefore little sensitivity to or tactical positioning on environmental policy issues; a consequent lack of 'green' political parties and strong, effective NGO efforts (until just prior to the political changes, in the case of the latter); surreptitious support for environmental concerns by the Young Communists, in an effort to limit central Party control over environmental information and activism; state-established investment and pricing schedules that paid little attention to environmental questions and, indeed, encouraged environmentally destructive choices (see below); lack of independent local governments with capacity to exercise autonomous choices regarding environmental issues; and the inevitable placement of issues like acidification into the context of political competition between larger national powers to the East and West.

Likewise, the political changes - resulting in a conservative coalition national government for the first four years of post-socialist governance, subsequently followed by a left-liberal coalition beginning in 1994 - have altered the setting for policy making in important respects. Among the shifts are:

- More open political processes and agitation for more participation and involvement on the part of NGOs, citizens, and others, including on issues of environmental policy
- Political competition, which creates some possibilities for political leaders to advance their careers and party interests by pursuing environmental questions (though this incentive is limited by the perceived economic precedence afforded economic issues, as explained later)
- A significant reduction of the role of the state in controlling investment choices and the management of key industrial sectors, with privatization and similar initiatives now well underway
- Establishment of the beginnings of independent local governments and, as a consequence, the initiation of both problems and opportunities for local involvement in policy making and execution on environmental questions, and
- A complex and dynamic set of challenges now facing decision makers in Hungary, many stemming from the multiple difficult adjustments necessary in the transition to a market economy.

## An Economy in Distress

The Hungarian economy has faced exceedingly difficult circumstances since the political changes of 1989 (see, for instance, OECD 1993). The Eastern markets, and the Council for Mutual Economic Assistance - COMECON - collapsed. The economy was poorly structured to take advantage of markets to the West. The country was further handicapped by a huge foreign debt (more than 21 billion US dollars). Although the decision to privatize much of the economy was widely accepted and resulted in the largest foreign investments in the region, and although many new businesses were founded, the picture on balance was bleak.

Bankruptcies escalated, particularly among firms that were forced to deal with market conditions for the first time. The privatization process experienced significant implementation difficulties (O'Toole 1994), and in the early years was executed without substantial attention to the environmental problems inherited from firms' earlier period under state control. Industrial output dropped by 30 percent by 1993 and then stabilized, and the Gross Domestic Product (GDP) fell by 20 percent (see Regional Environmental Center, 1994b, p.A5-6).

Unemployment escalated, from virtually nothing to 12 percent between 1990 and 1992 (see, for instance, United Kingdom 1995). With these difficulties came, likewise, cuts in environmental investment (Öko Rt. 1992). (Nevertheless, it is worth noting that there has been no correlation between production decreases in various sectors and the rate of environmental investments in these sectors - nor between investments and estimated environmental risks - at least as of 1992; see Öko Rt. 1992.)

Governmental estimates of total environmental investment in 1992 was 0.9 percent of a declining GDP. Investment for this purpose remained below one percent through the 1990s, although the official expectation is for environmental investments to reach 1.5 - 2 percent of GDP by 1998 (see Ministry for Environment and Regional Policy 1994a). Meanwhile, annual inflation rates of 20 to 25 percent have severely eroded the living standards of the unemployed and pensioners.

While the serious economic distress reduced some immediate measures of environmental difficulty (for instance, emission levels derived from industrial sources), the changes can hardly be viewed as long-term solutions to air pollution difficulties. In fact, the problematic economy has made it more difficult for those espousing environmental causes - whether within or without government - to be influential. In many circles, advocacy of environmental issues is viewed as antagonistic to economic, indeed social, welfare.

Newer approaches to environmental issues, particularly those founded on principles of sustainable development, are officially espoused, including in the nation's new Environmental Act of 1995. However, there is little evidence of sustainable development principles being converted into practical action through government. Instead, public policy, as interpreted via programs adopted and in place now, focuses almost exclusively on regulating 'end-of-the-pipe' emissions through punitive regulatory controls that are neither very punitive nor broadly and carefully implemented (see the treatment of policies and their implementation below). In addition, highly ambitious technical standards in force for several years have not been matched by effective execution - perhaps in part because the standards have been set so stringently.

These political and economic changes make the Hungarian context a dynamic, often stressful one for citizens and a complex one for political decision makers. The changes that have occurred or are underway in Hungary, nonetheless, suggest some reasons why this instance is particularly interesting from a research standpoint. Another justification for particular attention to Hungary is the relative underattention that has been devoted by the international scholarly community to environmental action, particularly policy efforts linked to international agreements, in Central and Eastern Europe.

## Significance of the Hungarian Case

Much of the debate and decision making regarding acidification in Europe has been focussed, understandably, on Western Europe - particularly on the Scandinavian countries and the nations of the European Union (EU). (Some limited exceptions that offer partial West/East comparison include Lammers 1991; Levy 1993; and McCormick 1989. A recent study of LRTAP implementation and effectiveness in Russia is Kotov and Nikitina, forthcoming.)

This emphasis has been appropriate, since Sweden and Norway have been policy leaders on this issue from the beginning - indeed, these countries (soon joined in an important fashion by the Federal Republic of Germany and the Netherlands in the early 1980s) are largely responsible for placing the issue on the international agenda in a forceful way.

And the nations of the European Union have been consequential in influencing both the overall quantity of airborne pollutants travelling long distances (including transboundary flows) and the extent and nature of international agreements on the subject. The nations of the EU, particularly the policy 'leaders' (see Keohane, Haas, and Levy, 1993, p.17) in this sector, have made significant moves at the national level and in international negotiations culminating in important regime features (see Levy 1993; 1995). Most of the nations responsible for large-scale emissions travelling across national boundaries are members of the EU, and most of the documented damage from acid rain inflicted within Europe has taken place within the EU countries.

So it is hardly surprising that this set of countries has been at the center of many political initiatives and debates, of research and scientific collaboration on the issues, and of technical and policy innovation in support of environmental objectives. (Considerable efforts in support of burden sharing between Western and Eastern countries have also been a significant element of cross-national experience.) Indeed, the regulations of the European Union itself are now an important set of components in the overall acidification question.

But, whereas many of the sources of transboundary transfers of pollutants are located in Western Europe and the nations of the EU, some are not. The most obvious case is Poland, which has been and remains a major 'offender'. Other nations of Central Europe, including Hungary, have contributed in significant measure to the emissions load and, to a lesser extent, to the transboundary problem.

Furthermore, a treatment focusing only on Western Europe would omit many of the signatories to the CLRTAP, the oldest and most significant arrangement encouraging the international control of acidification and related problems. Several LRTAP signatories are in Central and Eastern Europe.[1] An understanding of the links between national and international arenas in the field of acidification clearly cannot ignore these additional cases. Indeed, as is well known, the role of the East bloc was crucial in encouraging the continuance of international negotiations and cooperation on acidification during the period prior to the regime changes during 1989-1991 (see, for instance, Levy 1993).

The inclusion of a case like the Hungarian one (extending back into the era prior to the political changes), consequently, allows an opportunity to test in a state-socialist setting the generalizability of (or perhaps to suggest modifications in) ideas such as the demand-supply policy model around which some current research on domestic-international linkages is organized (see, for instance, Hanf, et al. 1996). In fact, the development of a separate analysis on this subject, drawing appropriately from the Hungarian instance and probing the utility of such concepts as political 'demand' in one-party states, might be a useful subject for future research. Such an investigation, aimed at critiquing or generalizing the analyses conducted in multiparty democratic systems, is not reported on in this book. The Hungarian case, nevertheless, could be used to frame and develop additional theoretical ideas for more systematic examination.

There are other reasons to focus on nations in Central Europe in a study such as this one. The strong desire on the part of the current Hungarian regime to achieve full EU membership suggests the possibility of informal influence by the EU on the domestic decision making within a nation that ostensibly has had no 'seat at the table' in EU acidification negotiations. Are there conditions under which the EU influences the creation and implementation of national policy in non-member states like Hungary? Do such subtle channels in turn generate further ramifications as nonmember preferences and positions influence in even more complex fashions the international agreements on acidification during the 'second generation' phase? Determining whether and to what extent these possibilities obtain could also be a primary focus for a later research in national settings like Hungary.

Reference to the 'current regime', above, suggests another reason why an investigation of a country like Hungary should be of interest. In a manner fairly representative among the countries of Central and Eastern Europe, Hungary has undergone a major transition in domestic regime characteristics during the period under investigation for this study. The replacement of the state-socialist system with a Western-oriented, liberal democratic constitution and governing coalition in 1990 altered profoundly the context of national policy making, international negotiations, and even implementation arrangements within Hungary in a rapid period of time. Some of these changes were summarized above, but more systematic examination of their impacts on efforts to deal with acidification are clearly warranted. Some such efforts have been a part of the current project. (Several other recent studies of environmental policy and politics in Central and Eastern Europe have been undertaken. See, for instance, Slocock 1996; Tang 1993; and several chapters in Victor, Raustiala, and Skolnikoff, forthcoming. Commisso and Hardi 1997 offer coverage of selected relevant Hungarian developments.)

Inclusion of Hungary among the cases examined offers the potential, therefore, to explore how fundamental regime changes might alter the orientation and institutional arrangements of national actors, introduce the prospect for different coalitions of actors to dominate choices regarding acidification as well as many other issues at the national level, and shift not only national policy on issues like acidification but also, perhaps, international agreements like CLRTAP.

Indeed, Hungary offers a particularly interesting case since the current government represents even a further shift in political fortunes during the time period of interest. With a left-oriented coalition replacing the more conservative one in control from 1990 until the elections of 1994, the prospect of even more complicated national-level changes emerges, with the concomitant potential to be influenced by and in turn have impacts upon the international acidification regime.

Nor is it merely a matter of 'change': dynamics of institutional alterations, with consequent reverberations in the realm of action for acidification (as well as other fields). The regime changes at the national level could be expected to effect certain kinds of openings to international influence.

Some of these can be explained in the following chapter and explored by implication in parts of the subsequent chapters.

## Note

1 Canada and the United States are also participants.

# 3 Domestic regime changes in Hungary and hypothesized influences on acidification

Hungary, like many other countries, confronts challenges arising from acidification. And Hungary has participated in the CLRTAP since its inception and has agreed to abide by all subsequently-adopted protocols. It is possible, therefore, that improvements on emissions levels for the targeted chemicals might be due to national commitments to uphold LRTAP agreements, conscious and deliberate policy making in support of such environmental issues, and effective implementation.

A general model of the implementation of international environmental agreements, such as that sketched by Jacobson and Weiss (1997, chapter 15), must of necessity include a broad range of variables. Among those often mentioned as potentially important are characteristics of the environmentally relevant activity involved, characteristics of the treaty or agreement, features of the international context (as a result of their review of many agreements in a number of countries recently, Jacobson and Weiss include the perceived equity and precision of the obligations, the reporting requirements imposed, roles provided for NGOs, sanctions, and incentives), as well as factors related to the country or countries involved. Following Jacobson and Weiss, one might include in this last-mentioned category such items as basic parameters (previous behavior concerning the subject of the treaty, history and culture, physical size and variation, and number of neighbors), 'fundamental' factors (the economy and political institutions), and 'proximate' factors (administrative capacity, leadership, NGOs, and knowledge and information).

In a study such as this one, concentrating on one set of closely related agreements in one country, most of the often-hypothesized influences cannot be systematically studied. The activity itself and most elements of the international context are relatively constant for LRTAP - although some shifts can be noted and analyzed in later chapters, and one particularly important international influence - that of the European Union - has operated at quite different levels during the period under consideration and is discussed in the present chapter. Even characteristics of the agreements are mostly constant, since the LRTAP protocols under review here were structured quite similarly during the period being considered. (Some recently considered additional protocols constitute departures in form and could be expected to result in somewhat different behavior.) The main expected sources of variation, other than the developing regime surrounding LRTAP itself, are domestic. But unlike in many countries, where domestic shifts are visible primarily in the 'proximate' factors mentioned above, Hungarian domestic changes also reach to fundamentals of the economic system and political institutions, at least for the period under examination. This fact suggests why the LRTAP in the Hungarian case can be expected to be unusual, and unusually difficult to analyze.

In part because of the multiple sets of changes underway currently in Hungary, fundamental domestic social forces could have impacts on acidification results. Certain of these other paths of influence can be hypothesized as assisting in the effort to control acidification, even if unintendedly; others might be expected to impede environmental programs aimed at controlling the targeted emissions.

This chapter outlines some of the principal channels of possible influence, aside from the straightforward path of good-faith support for international agreements combined with innovative national policy making and energetic implementation (that is, those hypothesized via the proximate factors listed briefly above). These additional hypothesized influences on acidification in Hungary are revisited near the conclusion of the study to assess what the available evidence shows.

The hypothesized additional influences outlined in this chapter fall into three categories:

- European integration and competing international environmental regimes
- Changes in the domestic political regime, and
- Property shifts combined with market forces (shorthand for a complex array of dynamics that might have multiple impacts on acidification results).

The three sets of forces are related but can be discussed seriatim.

**European Integration and Competing International Environmental Regimes**

The first possible additional influence on acidification, particularly relevant during the last several years, is related to the orientation of the new national regime to influences and events beyond the borders of Hungary. With the collapse of the Soviet bloc, new governments throughout the European region not only adopted the formal constitutional principles prevailing in much of the rest of Europe; they also explicitly proclaimed a reorientation of national interests and perspectives with openness toward the West. Nowhere is this clearer than in the strong desire expressed by the national governments of the region for full membership in the EU as a matter of high priority.

Hungary's own transition, discussed shortly, has been fully consistent with this generalization. With the choice to 'go European' comes an array of opportunities and constraints for central European countries: the requirement to adopt European Union policies into national legislation, the chance to receive financial assistance in making the transition fully effective (particularly through the EU's PHARE program), and so forth.

The earliest agreements derived within the LRTAP regime were enacted before there was any apparent possibility of Hungarian involvement in European integration; therefore the influence of the EU is not an issue for that period, as indicated later in this study. The matter becomes more important for the most recent few years. For this phase, it can be expected that the shift in domestic regimes may be accompanied by greater interest in and desire for compliance with EU-originated policies and preferences on many issues, acidification included. (For the symbolic and broadly political importance of integration with the EU in another country of the region - with import for environmental policy - see Slocock 1996.) Although some sectoral programs potentially supportive of EU objectives were established in earlier years, the increased interest in EU links within Hungary could be expected to attract additional political attention and financial resources to EU-oriented goals.

In fact, to the extent that resources are limited and the full array of international agreements on acidification requires a rationing of effort, the hypothesis can be framed more precisely: In Hungary (and neighboring countries), the regime transformation has been accompanied by national adoption and greater implementation of EU-originating international agreements on acidification (particularly the Large Combustion Plant Directive

(LCPD) which sets emission standards for nitrogen and sulfur from large power plants, and to a lesser extent the EU vehicle emissions standards), and also a shifting of attention toward sectoral regulations (prevailing in EU directives) at the expense of transboundary issues (including LRTAP agreements). This expectation is reasonable, even though the Hungarian direct contributions to CLRTAP have been primarily in-kind and staged in earlier years.

This hypothesis can be tested by changes in adoption and compliance within Hungary, by shifting patterns of implementation, and perhaps also by redistribution of resources and attention and even levels of international participation away from LRTAP and toward EU forums for international research, discussion, and negotiation of acidification agreements. While too little time may have elapsed thus far for a careful test in Hungary of this hypothesis, some preliminary assessment can be offered.

## Domestic Political Changes

The shift from a one-party regime to a democratic, competitive political system could also be expected to carry implications for acidification. As explained in Chapter 1, the old regime stifled political dissent and thus muffled potential public demands for more environmentally benign investments and decision making generally, at least until the years of the late 1980s, when environmental protest became increasingly popular (and tolerated) as a channel of political expression. Shortly before this latter time, scientific awareness within Hungary of the impact of environmental threats had increased considerably. As the political changes approached, environmental questions grew in public attention and discussion; but the environmental movement was also used opportunistically by politically ambitious individuals and groups to attract supporters for short-term advantage rather than long-term policy action.

What expectations for acidification might accompany the shift in domestic political systems? More openly competitive political and policy processes essentially place the issue of acidification, as well as other policy questions, into contention with other interests and forces. In a simple demand-supply model of an open and competitive political system (see Hanf et al. 1996), success on acidification can be expected to be a function of the balance of support for and opposition to reduced emissions.

No clear prediction can be made in the abstract, therefore. On the one hand, the change in regimes unleashes the possibility that environmental activists within and without government, NGOs, scientists and others could press the political system for more direct measures and aggressive control programs in the interest of environmental quality. On the other hand, acidification and other environmental issues face competition with other salient questions, including some - like economic health and social welfare protections - that may be perceived as directly antagonistic to action for environmental quality, at least in the short run. The challenge has been all the more difficult as national authorities have trimmed budgets for cultural activities and support of the increasingly responsible local governments.

One could expect, therefore, that the domestic political changes would mean that acidification would shift from an issue primarily under the control and in the hands of the central political elites to a matter subject to the more direct balancing of broader political forces. How the balance is struck is likely to be dependent on such competing issues as the strength of the nation's economy.

## Property Shifts and Market Forces

A third broad hypothesis deriving from the regime transformation in Hungary has to do with the choice for a mixed economy. This feature of the Hungarian post-socialist regime finds its policy expression in complex programs of denationalization (including property shifts to local governments and to selected institutions like churches) and privatization. It can be expected that massive shifts in property, like those initiated in Hungary during and even to a limited extent prior to the regime transformation, will have discernible impacts on policy and implementation for acidification.

In general, it can be hypothesized that denationalization and privatization have multiple effects for acidification policy nationally, and perhaps also Hungary's position in international negotiations. It is less clear what the overall impact can be expected to be, since some of the hypothesized links work in opposed directions (see also Hanf 1994b).

### *Institutional Differentiation*

One possible effect has to do with the institutional differentiation attending the property transfers: the more independent actors in possession of potentially-polluting properties - industrial plants, for instance - and the more actors involved in the implementation arrangements for Hungarian policy - newly-energized local and regional authorities, for example - the more complex and cumbersome the implementation efforts in the early years of regime transformation. (Note the link here to the idea that state targets can be more easily controlled; see Hanf et al. 1996.)

This expectation is really an amalgam of two:

- The impact of the sheer multiplicity of targets emerging from the property shifts in several directions (thus, too few regulators chasing too many targets, especially in an era of severe resource constraints for implementers); and
- The move toward private and away from state targets.

On the latter point, it may be useful to note that the hypothesis included in earlier research on international environmental agreements is plausible but contradicts some arguments in the literature on regulation (for instance, Wilson and Rachal 1977) to the effect that government agencies can be expected to have particular difficulties regulating other government entities. (See Durant 1985, for a careful treatment.) The issue, therefore, deserves careful empirical examination in the Hungarian setting.

*Issue Scope*

A second matter has to do with the likelihood of 'mismatch' between highly decentralized institutional arrangements, on the one hand, and the scope of decision making and coverage needed for effective control of acidification - particularly with respect to the transboundary issues - in any nation, on the other. As explained by Hanf and Underdal (1996), many environmental questions, including acidification, necessitate treatment on a scale sufficient to avoid suboptimal outcomes stemming from local or regional decision makers' undervaluing of the broader importance of their own negative externalities.

In Hungary there are 3200 local governments, and at the end of the 1980s these were charged with many substantial responsibilities. Before this time, their role was primarily to implement the decisions of the centralized regime. In a highly-decentralized institutional setting, the execution of acidification policy, whether nationally- or internationally-driven, is likely to be prone to implementation deficits (compare Scharpf,

Reissert, and Schnabel 1976, on the general issue of structure/policy mismatches and the tendency of overly-decentralized structures to exhibit 'control deficits').

This tendency might be particularly acute in countries like Hungary and on issues like environmental policy, during the period under investigation. The incentives for a competition toward environmental laxity, implied in the analysis of the preceding paragraph, can be increased still further when economic stagnation is an important part of the context, as it has been in Hungary for several years.

When local governments have large portions of their citizenry experiencing economic dislocation, unemployed, or threatened with loss of work, these units will have special reasons to treat broader, longer-term, and less salient issues like acidification relatively lightly. In the smaller cities of economically hard-pressed regions of the country, particularly where a large portion of the local economic industrial base has been dependent on one or a couple of state-owned industrial plants, this tendency can be expected to complicate implementation of policies aimed at control over acidification.

### *Inexperience and Limited Administrative Capacity*

Yet a third related barrier to smooth and effective implementation is that the multiple new governing actors, at the subnational level as well as centrally, are unlikely to be able to operate with sufficient expertise and administrative capacity to deal with complicated acidification matters with dispatch. This kind of factor, characterized as 'proximate' by Jacobson and Weiss (1997) in their general framework, can be expected to be rather deep-seated in a transitional regime like contemporary Hungary. Indeed, a broader aspect of this case analysis is an examination of the importance of administrative capacity, and institutional capacity more generally, in converting 'national will' into appropriately executed action.

In short, institutional differentiation, mismatches between the acidification policy problem and the considerable devolution of decision making for implementation to nonnational actors, and also the shortage of expertise and capacity for action in highly-decentralized systems all suggest the likelihood for substantial difficulties in executing whatever national policies are enacted in Hungary.

Still, these are not the only forces at work. And some additional considerations point toward the potential for enhanced control of acidification as a result of the property shifts and introduction of market forces.

### *'Adaptive' Implementation Arrangements*

One argument presented above suggested a possible mismatch between the acidification problem and institutional arrangements involving substantial devolution to small decision-making units. However, some scholarship on policy implementation also argues that, over the longer term, decentralized arrays offer certain advantages.

Ultimately, it might be expected that implementation of acidification policy will be improved by decentralization of formal authority (and development of subnational administrative capacity). Research on implementation suggests such a proposition, at least in circumstances when substantial 'local presence' (Porter 1976) or more 'adaptive' implementation patterns (Berman 1980) are warranted by the policy problem being addressed (see also Hanf, Hjern, and Porter 1978; Hull with Hjern 1987).

Institutional differentiation, then, can be expected to create implementation capacity that matters in the longer term, perhaps particularly at the point when differentiation is balanced by networking across interdependent actors in local government, ministries, industry, and among other interested parties like environmental NGOs. Unfortunately, this proposition cannot yet be tested, at least not longitudinally through this case, for the post-socialist Hungarian regime has only just begun its implementation efforts under the newer structural arrangements, and these patterns are even yet not firmly in place.

## Market forces: A Double-Edged Impact?

The shift to a market economy can be expected to have major changes on many aspects of life in Hungary, including acidification policy and its implementation. Interestingly, plausible arguments can be developed suggesting both that the market *constrains* what may be possible regarding acidification in Hungary, and also that it *encourages* reductions in air pollution (compare Slocock 1996 on this point, in his analysis of environmental policy in the Czech Republic).

### *Market as Constraint*

The hypothesized negative impact can be sketched. First it is useful to review economic developments in Hungary. The state-centered economic structure that was standard within the Eastern bloc for decades was liberalized somewhat in Hungary, beginning in approximately 1968. The nation entered its transition to a mixed economy with more experience in market-like settings and also more liberalized economic arrangements than neighboring formerly-socialist countries. Of course, this distinction has been an advantage in many respects.

Furthermore, on matters of environmental policy the state-centered economies of Central and Eastern Europe provided distorted incentives for investment. These resulted until the mid-1980s in heavy funding for environmentally unfriendly industries, technologies, and industrial processes (Bochniarz, et al. 1992). The result was the worst of both worlds: economic stagnation and also environmental degradation.

Despite Hungary's pragmatic experiments with liberalized economic institutions even before the transition period began, the nation did not escape the main features of the common economic-environmental plight of the region. Little could be said in the Hungarian case, then, for continuance of state-centered economic arrangements.

Nevertheless, when it comes to the implementation question, when international or national decisions are to be converted into stable patterns of action toward established, publicly-decided policy intentions, the shift to a market-centered economy entails substantial complications.

These stem primarily from the basic facts of life in market settings. As Charles Lindblom has bluntly put the issue, political choices framed in capitalistic settings are inevitably constrained by what he calls 'the market as prison' (Lindblom 1982; see also 1977). Once adopting market principles (even in mixed settings), governments find themselves having to induce the cooperation of business firms and business interests. Compliance cannot be compelled - at least not past a certain point. The foundational constraint is that businesses cannot be required to produce, or to continue in existence. To the extent that continued participation becomes unprofitable, business interests can and do cease operations. The need for some degree of 'voluntary' contributions on the part of for-profit enterprises in the relevant policy efforts in capitalistic settings means that the implementation challenge can be magnified in practical terms.

National regulations, no matter how carefully formulated or how urgently 'needed' in the eyes of the public or policy elites, must be sufficiently acceptable to economic actors so that they continue production of the goods and services produced by the economic order. Command and control is simply impossible.

Of course, it can be argued with justification that such top-down arrangements were largely fictions even in state-socialist systems. Indeed, as has often been documented in countries like Hungary, party politics and even personal cliques helped to determine the allocation of economic inputs in complex and often obscure bargaining processes during the days of ostensible central control.

However, the basic decision to shift to liberalized arrangements and market-driven economic decision making has meant a very large enhancement in the spheres for independent action possible within the economy. And with this shift, the implementation challenge for public officials, in particular, has grown. The task now is not to oversee and control, but to monitor and induce; protect the integrity and intent of public policy; and/but allow substantial operating flexibility to remain in the hands of private actors, including firms. More negotiation, more sophisticated reliance on a broader range of policy and implementation instruments - like incentive-based arrangements - is necessary. The economic shifts, then, and the attendant privatization, impose limits on public action while also increasing the possible reach of policies that are adopted.

The implementation complications deriving from the broad shift in the direction of privatization can be expected to be particularly acute in the nations of Central and Eastern Europe, which may be perceived as less stable and less immediately promising loci for investment. In other words, state action when the market is 'prison', as analyzed by Lindblom, is further constricted when the market is nascent. The need to attract capital, and the impossibility of compelling its risk - especially internationally -, means that governments in the transitional settings must offer especially favorable conditions for newly-forming private actors. Otherwise, further stagnation or even economic collapse can be expected.

### *Market as Stimulant for Environmental Improvement*

The foregoing logic, then, suggests market-grounded constraints on the potential impact of any international regimes for controlling acidification. However, there is another side to the market story, or hypothesis.

Despite the economic difficulties of the last several years, privatization is advancing in Hungary. There are now more than seven times as many private firms as there were in 1989, many of these being the product of new entrepreneurial initiatives rather than privatization, strictly speaking, of formerly state businesses. As the economic changes take hold, it can be expected that these shifts themselves - leaving aside impacts from national policies or international protocols - will have positive effects on the acidification problem. Redistribution of investment toward more prosperous and less energy intensive sectors is likely to bring a reduction at least in sulfur dioxide levels, even after the current economic downturn has run its course. Even now, 'the traditionally overdeveloped sectors like metallurgy and mining are declining, which is good news from an environmental standpoint. The building materials and chemical industries seem to have comparative advantages' (Regional Environmental Center, 1994c, p.37). More specifically on the energy issue:

> Since 1989 the energy consumption of the economy is on the decrease, in 1990-91 by 4.5% per annum, while communal and residential con-sumption shows a slight increase. The strong advance of that latter sector in energy consumption is really striking if one compares its consumption with the data of 1980. However, specific energy demand of industrial production increased at the same time. Namely while in 1991 industrial production decreased by 21.5%, its energy consumption only by 16.7% which means that we are still not exploiting the oppor-tunities of energy saving. [Hungarian Commission on Sustainable Development, 1994b, p.13]

Furthermore, the internationalization of the Hungarian economy introduces the prospect of participation by many more firms that have experience in other settings - including those with tougher environmental standards - and that can take advantage of technological advances to reduce emissions levels. Economic integration under market conditions thus holds the potential to improve the overall level of environmental impact by private firms, although some analysts worry about Hungary and other transitional nations becoming dumping grounds for outmoded technologies or foci of exploitation for environmentally-careless entrepreneurs.

Overall, then, it can be said that Hungary, being perceived as one of the most stable regimes in the region and demonstrating pragmatic interest in markets and incentive-based strategies since before the 1989 changes, is more favorably positioned than many other post-socialist nations. However, the balancing act between fostering a new market-based economy, on the one hand, and shifting policy to require much more attention to environmental issues, on the other, is a most difficult one here as well.

For all these reasons, then, the inclusion of a case like Hungary can add dimensions of interest to the current investigation and can stimulate exploration of additional research questions.

Of course, there can be too much of a good thing. Since nations like Hungary have undergone national regime transformations and a reorientation toward the EU at approximately the same time that second generation acidification agreements at the international level have been negotiated, any observable change at the national level has the potential to be explained by multiple causes. If Hungary meets its international commitments, is the reason the strength of the acidification regime, or perhaps the altered political setting, the opening to the European Union, the introduction of market forces, or the pattern of institutional differentiation and property shifts that have altered responsibilities in the transitional period? The problem confronted here is the classic one of overdetermination. However, an effort can be made to search for influences through these different causal paths at least somewhat independently; and to determine which, if any, of the dynamics associated with Hungarian transformations are related to which national (and possibly international) impacts.

The research reported here does not test all the foregoing hypotheses in a rigorous manner. However, in the field investigations and data analysis conducted for this investigation, systematic efforts were made to identify evidence bearing on all of these hypothesized lines of influence.

In the following chapters, the Hungarian experience on acidification is reported and the basic determinants of national positions, national responses to international agreements, and implementation efforts are treated. Emissions results are also included in the coverage. The final chapter reviews the central research themes and also revisits the additional hypotheses outlined above with an effort to draw some tentative conclusions.

The following chapter outlines some aspects of environmental quality and acidification in Hungary, including coverage of how the environmental issue is seen in contemporary domestic politics. Subsequent chapters turn to policy, policy implementation, and institutional arrangements for dealing with environmental questions, particularly regarding acidification. These subjects provide the context for a more detailed examination of Hungarian participation in and influence by the CLRTAP.

# 4 The environment, air pollution, and acidification in Hungary

This chapter presents an overview of how environmental issues have been considered and treated recently in Hungary, although coverage of tangible policy responses is omitted here. Policy is covered more carefully in Chapter 5. Attention is also directed in the present chapter to the severity of air pollution and acidification problems in contemporary Hungary.

## The Environment as a Domestic Issue

The emergence of concern with acidification in Hungary has been a relatively recent phenomenon. Under the earlier socialist regime the government was at first relatively unresponsive to the growth of public interest in environmental matters. Hungary did not participate with representation at the United Nations Conference in Stockholm in 1972, for instance, although the country did accept the recommendations developed at that meeting (Ministry for Environment and Regional Policy, 1991b, p.23). East-West political differences posed a barrier to broad international cooperation in the early 1970s. And, while many small steps were taken by fledgling nongovernmental organizations (NGOs) to encourage interest in environmental questions outside the direct supervision of the central Communist structure, the mass development of citizen involvement in pressing for the addressing of environmental issues awaited the Danube movement in 1984, which sought to halt construction of the Bős (Gabcikovo)-Nagymaros dam.

Even now, public opinion is 'ambivalent' regarding environmental matters generally. Concern is increasing, but 'the general level of ecological consciousness is still rather low' (Ministry for Environment and Regional Policy, 1991b, p.24). The economic stresses of the current period contribute to the restrained interest on the part of the broad public.

In recent public opinion surveys, the issues receiving the highest rankings in terms of salience have been those associated with the economy (inflation, unemployment, social welfare, and pensions). The environment has ranked in the middle of the list. While it may not be literally true, as one recent analysis reports, that the environment has 'fallen off the political agenda' (Regional Environmental Center, 1994b, p.A4), it is clear that there is little popular support for environmental measures that pose a risk to the fragile economy. This circumstance is expected by most analysts to persist for the foreseeable future.

On the other hand, some recent evidence indicates that environmental perceptions on the part of private corporate executives and managers in Hungary - among domestically-owned firms, companies owned by foreign investors, and also firms of mixed ownership - are virtually identical to the perceptions held by managers and executives in other parts of the world. 'If anything, Hungarian executives are slightly more sensitive to the importance of environmental issues and more strongly agree that companies are responsible for the environmental impacts of their products even after they leave the factory'. The Hungarian executives and managers differ somewhat from their counterparts elsewhere on the most appropriate means of addressing environmental issues that affect their companies. Interestingly, for instance, Hungarian business executives are more supportive of indirect regulatory measures and the use of incentives to encourage environmentally sound company decisions - this despite the general lack of such policies in the current Hungarian setting (Vastag, Rondinelli, and Kerekes 1994; see the next chapter for a review of policies).

Also, a comparison of Hungarian public opinion on some environmental issues with responses in several other countries suggests a rationale for the Hungarian perspective (see Table 4.1).

Hungarians, in short, are currently burdened with many serious problems and do not see the environment as most urgent. Nevertheless, many people recognize the criticality of the environmental situation. They are so pressed they feel they cannot become active themselves on the issue, and their experience with government over the decades suggests to many of them that such activism would be unlikely to have an impact. Overall, the picture suggests a sober but not unreasonable or unrealistic perspective (see Flaherty, Rappaport, and Hart, 1993, p.93).

Meanwhile, the highly visible environmental activism of the late 1980s, which was based in real public concern but was also used opportunistically

**Table 4.1**
**Cross-national Public Opinion on the Environment**

| % respondents with opinion listed | Hungary | Poland | West Germany | United Kingdom | United States |
|---|---|---|---|---|---|
| Environmental problems most imporant issue | <1 | 1 | 9 | 3 | 11 |
| Environmental problems in nation "very serious" | 52 | 66 | 67 | 36 | 51 |
| National environmental quality "fairly/very bad" | 72 | 88 | 42 | 36 | 46 |
| Environmental problems affecting respondent's health | 55 | 80 | 72 | 53 | 67 |
| Respondent active in an environmental group | 6 | 9 | 10 | 10 | 11 |
| Citizen groups seen as having an effect | 29 | 42 | 65 | 70 | 81 |

Source: Flaherty, Rappaport and Hart (1993: 9), from Dunlap, *The Health of the Planet Survey* (1972).

by both citizens and those with political ambitions as a vehicle for opposing the regime more generally, has given way to less overt and highly fragmented, but still significant, efforts by a huge variety of nongovernmental organizations (NGOs). Several hundred NGOs operate in the environmental sector currently in Hungary. Some of these are well known, a few internationally, while the large majority are small and narrowly focussed. Almost all are poorly funded, and competition for financial support from foreign or international sources is keen.

The environmental movement, as expressed through the NGO community, is not organized into a broad coalition, nor is it experienced at acquiring and analyzing information from government or other sources. The government, for its part, is in the early stages of learning how to interact productively and openly with environmental NGOs and others. Some officials in certain ministries have developed working relationships and a degree of mutual respect and trust with representatives of certain environmental NGOs. These links, however, are relatively few.

With respect to acidification issues, and the CLRTAP in particular, the relevant NGOs have thus far not involved themselves in the details of the international protocols or their execution within Hungary. Concerns about air quality are usually treated as matters of interest from a domestic perspective. And, while some NGOs are active in seeking to shift patterns of energy use (for example) and to engage government attention more vigorously in air quality and energy conservation, no NGO ties to the international acidification regime or counterparts in other CLRTAP countries have been important thus far in the Hungarian context. When international interest in acidification issues was in its early stages, NGOs in Hungary were unaware of the difficulties. As they became better informed, the groups nonetheless did not see how they might be able to become involved. And more recently, as Hungarian NGOs began to understand the real importance of the issues, the economic changes within the country altered the context in dramatic fashion and limited opportunities for stable, regular input. Most groups focusing on air quality have tended to emphasize local pollution rather than transboundary issues. Among the more active NGOs of this type are the Energy Club and the (Budapest) Air Working Group.

**Air Pollution and Acidification in Hungary**

A few basic elements of data on Hungarian air pollution and acidification can be introduced here. Air pollution is significant in a relatively small area of the country - estimated in the early 1990s variously at approximately eight (Bochniarz, et al., 1992, p.18) to 11 percent of the surface (Hungarian Commission on Sustainable Development, 1994b, p.26; Ministry for Environment and Regional Policy, 1991b, p.43). Approximately 40 to 50 percent of the nation's population was estimated to be regularly exposed to serious air pollution threats (Bochniarz, et al., 1992, p.18; Bulla 1992; Ministry for Environment and Regional Policy, 1991b, p.43).

The most recent official estimates of the severity of the air pollution challenge in Hungary conclude that no large regions of polluted air remain in the country. Less than four percent of the territory can now be classified as polluted, while 9.3 percent qualifies as moderately polluted. These areas, however, affect approximately 52 percent of the population, an even larger segment than before (Environmental Council of the Parliament 1995).

The most serious concerns have been concentrated in major population centers and the vicinities of large industry, particularly with respect to human health issues. Still, as the treatment below shows, long-range transport into and out of Hungary, as well as within the nation, are also matters of significance. (Hungarian officials have concluded that relatively discrete, locally concentrated emission sources are a major source of transboundary flows of airborne pollutants.)

Measures of emission of major air pollutants, including those that are the foci of international agreements, suggest that in ecological-scientific terms, acidification is a relatively important matter in Hungary, although as an environmental issue it is surpassed by other pressing problems. This statement is supported by data depicting the Hungarian situation in both relative and absolute terms.

Regarding the former, for instance, at the start of this decade, Hungary ranked fourth in the dubious competition among nations of Central and Eastern Europe for the highest quantity of sulfur dioxide emissions per capita. The national standing for nitrogen oxides in the region, again based on a per capita measure, was fifth or sixth.

In the case of sulfur dioxide, one contributing influence has been the fact that domestic sources of coal are high in sulfur content. As discussed later, the levels of sulfur dioxide did decline through the early 1980s - primarily because of shifts to other sources of fuel prompted by other considerations, and not mainly because of environmental reasons - and have continued downward since then, more due to economic difficulties than to the implementation of policy in favor of a cleaner environment.

Perhaps an even more noteworthy measure is one standardizing emissions for level of economic activity, since such a calculation takes explicit account of the environmental-economic duality lying at the heart of the goal of sustainable development, a policy objective embraced in the formal pronouncements of the Hungarian government as in many other countries. As one report puts the matter:

> Emission of both pollutants per $1000 of [Gross National Product] in early 1990 was almost 9 times higher than the average in the countries which make up the European Community in the West. The comparison of indices of productivity demonstrates the serious gap between Hungary and the most developed countries of Western Europe in terms of technologies and resources available for investment in environmental protection and restructuring. In short, the low levels of per capita productivity in Hungary, coupled with the high levels of air pollution per unit of productivity creates a severe obstacle for investment. [Bochniarz, et al., 1992, p.18]

This particularly unfavorable situation reflects in significant measure the choices, and miscalculations, made by the government prior to the 1989 transition to a liberal regime with a goal of a market-based economy. Unlike most Western European countries, Hungary's centrally-planned industrial base was built upon a set of assumptions that have come back to haunt decision makers today. From the 1970s until nearly the time of regime transformation, the nation embarked on a set of industrial investments based on an expectation that the economic future lay in large mass-production enterprises. This choice placed Hungary behind most of its Western neighbors, who were moving toward service- and knowledge-based economies.

The investments also concentrated on energy-intensive industries, thus placing further burdens on both the economy and the environment. In fact, the political regime treated the oil price increases of the 1970s as aberrations. Instead of adjusting economic planning away from energy consumption in view of an expectation of higher energy prices over the longer term, the country chose instead to safeguard, even increase, energy supplies and production (Ministry for Environment and Regional Policy, 1991b, p.11-13).

As a consequence, Hungary entered its transitional period toward a democratic political system and market economy with great reliance on heavy industry (see Kerekes, 1993, pp.140-41) and little investment in energy-saving and environmentally-friendly technologies. Currently, Hungary uses energy at approximately twice the intensity of Western Europe, although Hungarian energy efficiency is one of the highest among the nations of Central and Eastern Europe (Hungarian Commission on Sustainable Development, 1994a, p.11-12).

The ill effects of these choices were further exacerbated by the centrally-planned suboptimal allocations of investments over time, particularly the underinvestment in essential infrastructure - including environmental infrastructure - and environmentally threatening concentrations of polluting facilities in the same vicinities (see Ministry for Environment and Regional Policy, 1991b, pp.16-18).

Relatively low investment in the environment has been reflected in the modest ratio of environmental inputs in the economy to the GDP, as explained earlier. In 1980 the former was 0.7 percent of the latter; over the succeeding decade, the peak for environmental inputs, relatively speaking, was 1.0 percent. In 1990, the ratio was 0.7 percent once again (Ministry for Environment and Regional Policy, 1991b, p.20).

In part because of the topography of the nation - with approximately three-quarters of the surface consisting of flatlands lying only a couple of hundred meters or less above sea level, and with the nation mostly comprising the lowest portion of the Carpathian basin - transboundary air pollution is quite significant, although the absolute quantities are not as sizable as for larger and more industrialized nations. Transboundary flows are important for both import and export. The great majority of Hungarian emissions travel across national boundaries before deposition (Flaherty, Rappaport, and Hart, 1993, p.12). And winds, prevailing from different directions in different regions of the country, bring quantities of airborne sulfur dioxide and nitrogen oxides sufficient to comprise almost half of the country's total deposits from the air (Ministry for Environment and Regional Policy, 1991b, p.4).

Tables 4.2 and 4.3 show levels of emissions in Hungary, over time, for two key air pollutants closely linked to acidification and international regimes for control of the problem: sulfur dioxide and nitrogen oxides.

**Table 4.2**
**The Trend of Sulfur Dioxide Emissions in Hungary (Kilotons/year)**

| Sector | 1980 | 1985 | 1990 | 1991 | 1992 | 1993 |
|---|---|---|---|---|---|---|
| Households | 290.6 | 303.5 | 222.0 | 209.2 | 127.6 | 127.4 |
| Services | 44.9 | 36.7 | 29.0 | 34.2 | 27.0 | 18.0 |
| Transport | 49.0 | 21.1 | 16.0 | 13.4 | 12.9 | 7.6 |
| Electricity generation | 654.7 | 504.0 | 423.0 | 407.5 | 442.7 | 426.8 |
| Other heat production | 33.3 | 21.9 | 12.0 | 13.2 | 13.7 | 15.5 |
| Industry | 522.2 | 487.3 | 286.0 | 219.4 | 192.4 | 149.8 |
| Agriculture | 38.1 | 29.1 | 22.0 | 16.0 | 11.0 | 12.2 |
| **Total** | 1632.8 | 1403.6 | 1010.0 | 912.9 | 827.3 | 757.3 |

Source: Ministry for Environment and Regional Policy, Government of Hungary, 1995.

Estimates of quantities of volatile organic compounds (VOCs) are more difficult to generate. Available estimates covering several years for nonmethane VOCs (NMVOCs), however, are listed in Table 4.4. The estimates and trend data are presented by sector in these displays, thus facilitating later discussion of explanations for the observed changes over time.

For now it can be said simply, and not surprisingly given the nature of the processes linked to such pollution generation, that the major activities resulting in high quantities of $SO_2$ emissions have been energy generation and heavy industrial processes. Similarly, it is easy to see that transportation has been the primary contributor to nitrogen oxides ($NO_x$) emissions. Solvents and transportation are, again not surprisingly, dominant as sources of NMVOCs.

**Table 4.3**
**The Trend of Nitrogen Oxides Emissions in Hungary (Kilotons/year)**

| Sector | 1980 | 1985 | 1990 | 1991 | 1992 | 1993 |
|---|---|---|---|---|---|---|
| Households | 18.2 | 21.5 | 19.0 | 18.9 | 14.9 | 14.5 |
| Services | 7.1 | 7.7 | 7.0 | 6.4 | 4.8 | 4.8 |
| Transport | 111.3 | 110.5 | 116.0 | 98.4 | 94.3 | 91.7 |
| Electricity generation | 69.0 | 61.6 | 45.0 | 35.3 | 37.3 | 41.3 |
| Other heat production | 4.1 | 3.8 | 3.0 | 3.5 | 3.7 | 4.2 |
| Industry | 53.3 | 48.8 | 41.0 | 35.6 | 25.0 | 24.1 |
| Agriculture | 9.9 | 8.6 | 7.0 | 5.0 | 3.2 | 3.4 |
| **Total** | 272.9 | 262.5 | 238.0 | 203.1 | 183.2 | 184.0 |

Source: Ministry for Environment and Regional Policy, Government of Hungary, 1995.

The data confirm quantitatively what was stated earlier regarding trends over time. For sulfur dioxide in particular, an earlier downward tendency in emissions of this pollutant has continued in the 1990s. It is clear from economic data and interviews with a set of the relevant actors, however, that these more recent reductions are mostly epiphenomena of the wrenching economic contractions being experienced in Hungary during the last several years. An additional element has been the shift to some reliance on nuclear power, a move initiated by government but not in response to international agreements (see Öko Rt. 1992).

The trend data on sulfur dioxide emissions for several separate kinds of sources during recent years are displayed in Figure 4.1 for a slightly different time period (see also Ministry for Environment and Regional Policy 1994b). These data show visually the especially important contri-

**Table 4.4**
**The Trend of Emissions of NMVOC (Kilotons/year)**

| Emission Sources | 1988 | 1991 | 1992 | 1993 |
|---|---|---|---|---|
| Energy sector and district heating | 1.0 | 1.0 | 0.9 | 1.0 |
| Communal heating | 25.0 | 18.5 | 18.0 | 17.0 |
| Industrial heating and technologies | 10.0 | 6.5 | 6.0 | 6.0 |
| Use of solvents | 78.5 | 45.0 | 43.0 | 46.0 |
| Losses in transportation, primary and secondary distribution | 90.5 | 72.5 | 67.9 | 73.0 |
| **Total** | **205.0** | **143.5** | **135.8** | **143.0** |

Source: Ministry for Environment and Regional Policy, Government of Hungary, 1995.

bution of the industrial slowdown to the overall emissions reductions.

In the case of nitrogen oxides, the figures reflect less dramatic changes. The trend in the last few years has been modestly downward. In part this outcome is due to the economic difficulties of recent years (see Table 4.3), and in part the short term trend reflects some gains in the transportation sector. Here there have been two counteracting trends: (1) replacement of many old, poorly-maintained, heavily polluting motor vehicles with newer, better equipped, and more environmentally friendly vehicles from Europe, Japan, and the United States; and (2) an increase in total numbers of vehicles on the roads. As the shift to less polluting vehicles becomes more complete in the next few years (through obsolescence and also via modest support from the government for shifting to less polluting personal vehicles), it can be expected that the transit sector will become a growing contributor to the $NO_x$ load.

Hungary now has fewer vehicles than is the norm in Western Europe (approximately 200 per 1000 people vs. 350 per thousand), but recent

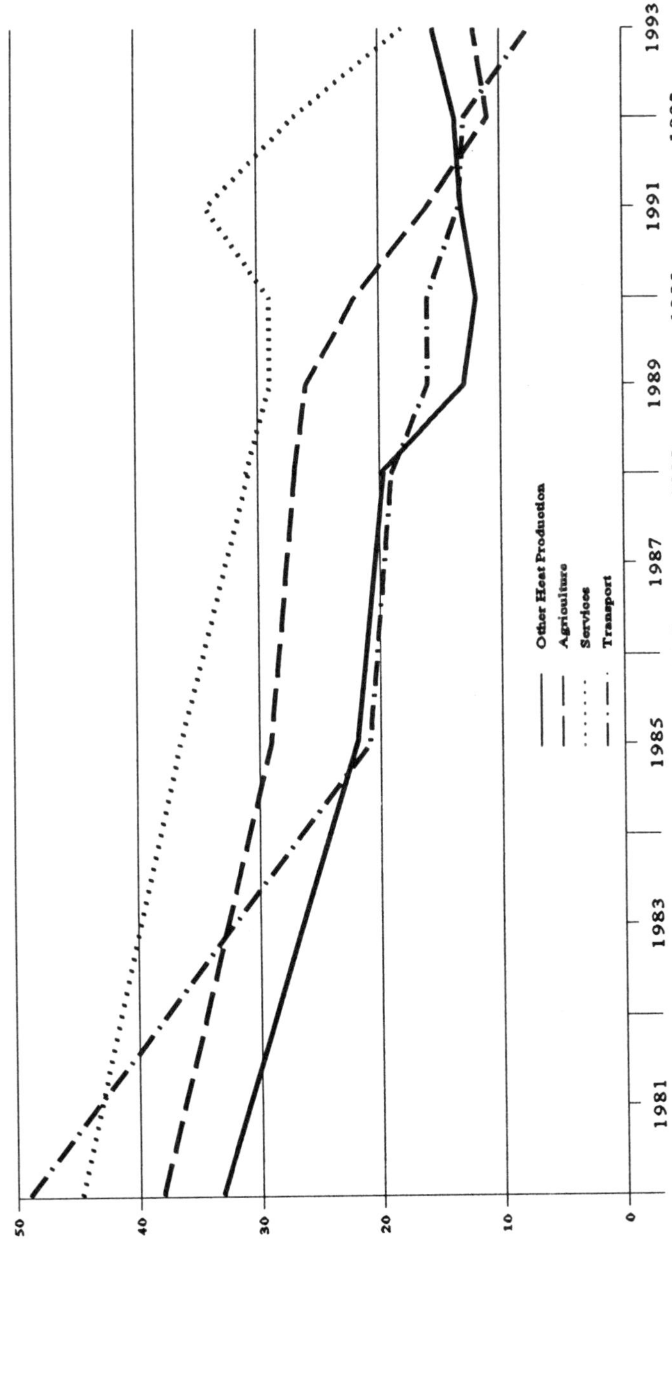


**Figure 4.1: Emissions of Sulfur Dioxide in Hungary (in Kilotons/year)**

years have seen a sudden inflow of cars produced in the West (Kaderják, 1993, p.3), as public demand for increased mobility has grown. It is projected that as Hungarian industry develops lighter industrial subsectors, a more developed roadway system will be utilized heavily. Current plans of the Ministry of Transportation, Telecommunication and Water Management call for a 'massive road transportation sector in a future Hungary' (Kaderják, 1993, p.5). Even now, however, the Hungarian government views air pollution from vehicles in urban areas as a 'serious problem' (United Kingdom, 1995, p.15).

The impact of acidification can be seen not only in the cities of the country; sulfur dioxide and nitrogen oxide levels in the air have been problematic throughout the country, at least until the last couple of years. Measurements of background concentrations of these pollutants 'increased dramatically' during the 1980s, and the pH of precipitation 'gradually decreased' (Ministry for Environment and Regional Policy, 1991b, p.48). Concentrations have been relatively constant since the mid-1980s.

> In terms of continental background pollution . . . Hungary belongs to the average countries in Europe. About 50 percent of the atmospheric sulfur and nitrogen compounds precipitating in Hungary come from abroad: [the Czech Republic, Slovakia], the former GDR [German Democratic Republic, or East Germany], Poland, and Italy. [Note: the portion of nitrogen oxides emanating from abroad is approximately 80 percent as of 1995.] About 27 percent of domestically emitted air-polluting substances precipitate in Hungary and 50 percent in the neighboring countries. In Hungary, atmospheric acid precipitation is also of average degree. The pH value of precipitation shows a declining trend. Between 1987 and 1986, its average amounted to 4.8, reflecting a higher degree of acidification than in the southern hemisphere where air is cleaner. The quantity of sulfur and oxidized nitrogen compounds precipitated in Hungarian forest soils susceptible to acidification is, for the time being, insignificant. [Ministry for Environment and Regional Policy, 1991b, p.48]

The quality of air in Hungary, discussed in abbreviated fashion above, is dependent not only on natural conditions and domestic emissions but also on transboundary inflows and outflows. The data on the transboundary flows in recent years show that Hungary receives considerable long-range pollution from abroad (see above) and also contributes its own share to the pollution from abroad (see above) and also contributes its own share to the problem.

Air pollution has obvious and widely-noted impacts in Hungary. As mentioned earlier, the most serious incidents have declined over time - particularly in Budapest, where during one three day period in January 1970 sulfur dioxide concentrations were measured as high as 3500-4500 tg per cubic meter. However, impacts of acidification remain significant.

The effects of acidification in Hungary include measurable human health consequences, particularly regarding respiratory ailments.

> Due to the health-damaging effects of air-polluting substances, acute respiratory diseases are on the increase. In settlements with polluted air, the rate of chronic bronchitis among the adult population is three times above the average, while the rate of bronchitis is gradually increasing among the age-groups between 7 and 17 years. Between 1979 and 1987, the occurrence of respiratory asthma more than doubled in Budapest. The capital's population, especially the student one, is frequently attacked by anaemia and osteogenesis disorders. The number of deaths caused by tumour is on the increase while children catch diseases four times more often than in moral [*sic*] conditions. [Ministry for Environment and Regional Policy, 1991b, p.49]

Overall, Hungary's population is relatively unhealthy by European standards. Life expectancy for men is approximately 65 years (and falling), for women 74 years. Experts attribute the 'alarming decline in the population's physical condition' to 'unsolved economic and social problems', including environmental conditions. 'As for environmental factors, it is air pollution that affects physical condition to the largest extent' (Ministry for Environment and Regional Policy, 1991b, p.21).

Deterioration of forests can also be traced in part to acidification.

By one expert's calculation, 'the damages caused by air pollution even by modest estimate is [*sic*] beyond an annual cost of 15 billion HUF [Hungarian forints]' (Bulla, 1992, pp.69-70). As the government indicates, 'according to conservative estimates, the yearly health damages (including costs and losses) caused by air pollution are at least three times higher than air protection inputs' (1992, p.23).

As a consequence of the evidence, Hungary's obligations under the LRTAP, and the country's status as associate member of the EU, the nation has committed itself formally to deal with these international dimensions of the acidification problem. Yet the delicate balance that such a policy requires under present economic conditions, in particular, is nicely captured in the language of a government statement:

> Owing to its geographical, political and economic position, external factors greatly influence the way Hungary handles environmental problems. As environmental pollution respects no boundaries, international efforts have to be coordinated. It is in our national interest both to cooperate with our neighbours on environmental matters in a regulated way and to support multilateral agreements controlling environmental effects both on a regional and a global level. Consequently, Hungary takes a lead in initiating improvement of environmental conditions in the Mid-East European region and contributes to dissemination of environmental information, research findings and technologies in this part of the world. Our cooperation with our neighbours is based on the realization that Central Europe should be regarded as a single ecological entity.
>
> As for our integration into the European region, a great deal depends on how much Hungary will environmentally be able to meet the challenges set by the single European market emerging . . . That is why we intend to introduce the environmental standards and norms of the European Economic Community gradually. [Ministry for Environment and Regional Policy, 1991b, pp.85-86]

How the government has responded to the environmental issue generally through the adoption and enforcement of policy is the subject of the next chapter.

# 5 Hungarian environmental policy and its enforcement

The general policy approach taken by the government of Hungary during the most recent couple of decades is the focus of this chapter.

## General Overview

Leaving aside governmental policies enacted as early as the nineteenth century and designed primarily to provide some protection against industrial hazards, Hungarian policy on the environment began in the 1960s and 1970s, with a series of laws enacted by the state socialist regime to cover a range of issues. Besides the law on the protection of air purity (1973), however, these were largely aimed at preserving resources important to processes of production.

For environmental policy in general, the most important piece of legislation adopted under the state socialist system prevailing until 1989 was the Act on Environmental Protection, enacted in 1976. This was the first explicit establishment of principles of environmental policy. The law established the (formal) right of citizens to live in a healthy environment and proclaimed protection of the human environment a responsibility of the broad society. Furthermore, regulations stemming from the 1976 Act were quite stringent, at least on paper. Nevertheless, as indicated below, they have been unsatisfactory in dealing with the practicalities of pollution, including acidification and air pollution generally (Hinrichsen and Enyedi 1990).

The political and economic changes of 1989 spawned constitutional change and new elections in 1990. The newly-elected coalition government announced in that year a *Program for Transition and Development of the Hungarian Economy* which contained discussion of environmental difficulties but proposed no specific program to address the issues. Two years later the government assumed sole authority for developing environmental plans; 'no parliamentary or outside body was permitted to develop an environmental program or policy' (Bándi, 1993, p.5).

Until late 1995, there remained no broad, comprehensive environmental policy in Hungary. The 1976 law, supplemented by numerous narrower pieces of legislation, several amendments to the 1976 Act, and many regulations developed in the interim, was the primary policy formally established in the country. The political regime had changed, therefore, and even the economic system had been dramatically altered, but the formal policy on environment had been held essentially constant.

The government's environmental action plan, prepared in 1993, stipulated a number of principles and priorities. However, these were not ordered or ranked in terms of priority, nor were they connected to programs and budgetary decision making. Furthermore, the programs and initiatives of other ministries - aside from the Ministry for Environment and Regional Policy - are not integrated with the action plan and are often inconsistent in practice with its objectives. Even the environmental ministry itself experiences difficulties articulating clearly its own priorities.

Similarly, the Hungarian Commission on Sustainable Development was created in 1993 'for the coordination of the national efforts in use of principles and implementation of concrete tasks of sustainable development, the efficient application of environmental considerations in various long-term sectoral plans, the increase of public awareness on relations of the development [*sic*] and environmental issues, or the environmental hazards [*sic*]' (Hungarian Commission on Sustainable Development, 1994b, p.3).

However, even now there are no national programs incorporating principles of sustainable development into the routines and regular decisions of the government or implementers of environmental efforts. The most visible signs of acceptance of such principles has been the adoption of a governmental resolution on April 2, 1993 regarding a commitment to follow up on United Nations Conference on Environment and Development (UNCED) sustainable development beginnings by incorporating such priorities into actions within the country (Decree of the Hungarian Government No. 1024/1993), as well as the acceptance of the new environmental policy in 1995.

It must be emphasized that the extent to which such formal commitments will influence the concrete actions and programs of Hungary remains to be seen. Even the national Commission admitted as much: 'since the United Nations Conference on Environment and Development, in fact, modest results have been achieved for this period at our national level in course of implementation of tasks accepted during the Conference' (1994b, p.10). This statement certainly holds for air pollution actions, as sketched later, despite the obvious interdependence of economic development and environmental quality in this subsector.

Indeed, evidence since the political changes suggests what one participant in Hungarian environmental initiatives terms 'hard fighting' between a number of economic experts, including several within government, and environmental advocates within and without government.

The former, along with some politicians, assert that there is currently not enough income for the government to initiate the technical changes needed to improve environmental conditions. The argument here is to opt for the strategy of focusing on economic growth now and using the income generated to convert technology, as necessary. Those advocating a more environmentally-focussed agenda argue that the time is ripe to eliminate archaic technologies and industrial operations, and to begin to alter consumer/household actions as well. In the three most environmentally-central sectors - energy, transportation, and agriculture - the government's efforts should be directed toward change as a top priority. And, say these advocates, the present time is ideal, since the economic transformations now create opportunities to undertake the conversion before major new investments are sunk into place.

This dispute, which occupied the first transitional government throughout its tenure, has not only generated a relatively indecisive governmental effort overall but has also provided entre to certain interests from abroad to exploit the weak governmental position for private gain or the exportation of environmental problems from other countries. (Some from abroad offer to provide new incineration technologies, for instance, in exchange for Hungary's willingness to receive solid wastes for ultimate disposal.) Some of those critical of the apparently vacillating stand of the Hungarian government on the economy-environment issues suggest that support from Western Europe could be most useful, not only in terms of financial assistance but also via national policies elsewhere to make it more difficult for Hungary to be exploited under the current economic conditions (for instance, on the waste disposal issue).

The elections of 1994 resulted in a new coalition government consisting of parties in opposition during the first four years after the political changes. A few comments can be offered regarding the perspective and approach of the newer coalition toward environmental issues.

First, in the view of most of the actors in the system, the new government is not behaving in distinctively different fashions on environmental questions, when compared to the previous coalition. (The newer coalition has brought back into decision making roles some of those formerly involved in the energy sector, a move viewed as helpful by some environmentalists.) Second, the new government seems somewhat more interested in ensuring that the processes of privatization and liquidation do not result in the ignoring of environmental damages or liabilities (see below). To that end, programs have been initiated to deal with some of the environmental aspects of the privatization process (the problem itself is complex; see Abel et al. 1993). Third, agencies and programs focusing on environmental issues face budgetary stringency, as the new government struggles to deal with the nation's serious economic problems. In fact, the environmental ministry itself has found its budget severely restricted and has had to deal with hiring freezes under the current coalition. Yet, fourth, the new government's policy statement on the environment contains a goal of 1.5 to 2 percent of the nation's GDP to be devoted to environmental inputs by the year 2000. Of course, this last-mentioned objective is merely a goal, but it is an ambitious one. As mentioned earlier, the nation's environmental inputs have averaged roughly 0.7 percent in recent years. In this connection, it is worth noting that the nation's Central Environmental Fund, discussed below, had achieved sizable proportions by early 1996 (approximately HUF 12 billion) and offers the potential for substantial environmental investments in the coming period.

Finally, it should be mentioned that national policy on environmental matters has recently begun to be modified. New legislation accepted in Parliament in September 1995 aims to begin a process of updating and systematically revising Hungarian approaches to environmental issues. Many aspects of the new legislation could eventually be significant. The broad endorsement of the use of economic instruments where feasible, as well as the explicit adoption of some important, if vaguely formulated, environmental principles are among the features of interest, as the legislation and other recent governmental statements are given concrete meaning in the future.

Furthermore, this legislation explicitly states that the country will address issues of transboundary pollution, even in situations for which there is as yet no explicit international agreement (item 11, 1995). Similarly, the policy would obligate the government to establish information systems sufficient to help fulfill international data submission requirements and provide strong support for efforts at international environmental cooperation (item 49, paragraph 2). Other supportive goals are endorsed. It may be, therefore, that these recent initiatives signal the beginning of a process in which the nation will enhance its efforts and results on matters closely related to the objectives of the CLRTAP, as well as on other international implications and environmental activities more generally.

Nevertheless, these broad pledges are as yet only statements of principle, not detailed programs of action that the government has adopted and funded, let alone implemented. Accordingly, attention is directed in this chapter to the operations of the Hungarian policy framework in place during the earlier regime and the present one, prior to the recently-accepted changes.

## Enforcement Overview

Standards for protection of the environment are central elements of Hungarian national policy, but serious problems of enforcement have long plagued the system. As one study reported recently, regarding most of the period covered by this investigation:

> Hungary's environmental plight also has roots in its legal and political system. The State focused on technical issues, rather than regulation and enforcement. As a result, monitoring, largely conducted by industry with poor equipment and very little regulatory oversight, produced little useful information. What enforcement and implementation there was tended to focus on sanctioning those who polluted rather than on preventing the pollution. When a penalty was imposed, it was often insufficient to provide an incentive to curb pollution. Perhaps most importantly, *there was no public participation in the decisionmaking* process. The role of civil law and the courts was very limited. . . . Even when citizen intervention was permitted, non-governmental organizations were too weak to support successful public involvement.

> *Organizational weaknesses* also limited Hungary's ability to protect the environment. The central environmental agency is much weaker than the economic ministries, both in terms of power and resources. Spheres of authority are poorly defined and are rapidly changing. This unclear division of power creates uncertainty and hesitation, not only for the environmental agencies, but also for the regulated community. [Bándi, 1993 p.2, emphasis in original]

Several of these issues, especially those regarding organizational questions, are treated in more detail later. Others, such as the problem of non-participation, are mostly a legacy of the one-party system and may be in the process of change. Here some broad observations about policy can be made.

Environmental policy in Hungary from the days of state socialism through the period covered by this study has emphasized punishment for violation of regulatory standards rather than other types of policy instruments. In particular, designing policies to prevent pollution in the first place has been neglected. A permitting process is activated prior to construction projects, but thus far comprehensive environmental coverage is not provided. The heart of environmental protection remains with the system of standards and associated penalties.

As one Hungarian environmental economist has explained, 'A strange paradox of our development is that Hungarian environmental protection regulations are in many respects stricter than the average [European Union] standards and much stricter than justified by our economic development level. The standards are so strict, in fact, that industries cannot comply with them' (Kerekes, 1993, p.146).

Some non-punitive efforts have recently begun with the aim of focusing more on preventive aspects of policy. For instance, during the first government after the political changes taking place in 1989, a governmental process for environmental impact assessment was adopted, and implementation has begun.

But in most spheres, non-punitive and especially extra-regulatory instruments are only sporadically employed. For example, economic policy instruments have been discussed but have been mostly notable for their absence in functioning programs. A superficial exception is the system of penalties for violation of standards, which actually functions as a fund-raising mechanism and not as a deterrent to pollute. (The Central

Environmental Fund may be evolving into a potentially important tool for encouraging the initiation of environmentally-beneficial efforts.)

In addition, environmental education is not included in the curricula of most schools, and the broad public has had only limited information about environmental issues. Forums for discussion of these matters and for public and NGO involvement in environmentally-important public decisions have been absent for the most part. The same is true regarding collaborative efforts with businesses in working to achieve the simultaneous goals of economic prosperity and also environmental enhancement.

And the thousands of new local governments, potentially important decision makers for a range of environmental issues, are in need of technical assistance on environmental questions as they undertake their responsibilities. This matter is of particular importance, since - as the discussion earlier points out - the local units are under heavy pressure to find ways of attracting economic activity in their jurisdictions and thus may have strong incentives to ignore environmental questions, at least in the short- and medium term (see Bochniarz, et al., 1992, pp.53-54).

The local governments' role is potentially important. However, before covering this topic, this volume turns to a review of current national air pollution policy in Hungary.

# 6 Hungarian air pollution policy

The policy strategy adopted by the government of Hungary to deal with air pollution issues for the period covered by this study has been largely consistent with the broad depiction presented in Chapter 5. The present chapter presents some of the principal aspects of Hungarian air pollution policy and also explains the situation of the local governments for the period since the political changes.

Following this exposition, Chapters 7 and 8 concentrate on the complex issues of organization and implementation (for the former) and institutional capacity and development (for the latter). These necessary treatments complete the framing of the context within which Hungary's participation in the acidification regime can be understood. Accordingly, Chapters 9 through 11 focus squarely on the CLRTAP and Hungary's involvement and accomplishments.

## Central Elements

Acidification issues in Hungary are dealt with most directly through environmental policy and programs aimed at controlling air pollution through formally coercive and regulatory means.

The punitive aspect of Hungarian policy can be seen by examining air pollution controls, a subsector fairly typical of environmental regulation in present-day Hungary. (Water regulations are more detailed than those for air.) Here, as with other sectors, the differences between formal goals and instruments, on the one hand, and actual execution, on the other, is great.

The 1976 national policy continues to establish the general approach taken for air and other aspects of pollution control. The system is reliant on German perspectives as they had been developed at the time of the early 1970s. (There are Swedish-style elements as well, since particular branches of industry are assigned specific emissions standards.) For air, the regulation was altered in 1986 and again in 1989. (A few changes initiated since the political changes of 1989 are noted in the discussion below.) The law establishes fines for environmentally-damaging activities, and the penalties are set on a schedule based on the extent and degree of hazard of the pollution. Repeat offenders are subject to heavier financial penalties.

Ambient air quality standards have been established for hundreds of materials in Hungary, and in this respect national policy appears to be quite restrictive. However, the permitting system and fine-based strategy for enforcement - especially when combined with the practical limitations of monitoring and enforcement through the regional environmental inspectorates - suggest a somewhat more permissive picture (see discussions in Flaherty, Rappaport, and Hart, 1993, pp.20-21; Várkonyi and Kiss 1990).

The overall system of air pollution control relies on emissions standards which divide all parts of Hungary into one of three kinds of air quality zones. The main focus of regulation is on stationary sources; some controls through product licensing are also included. The regional environmental inspectorates are empowered to stop or limit polluting activities, if such actions are deemed necessary to enforce the standards.

Especially serious situations may trigger exceptional measures to be taken, such as prohibition of the use of automobiles or stipulation of the use of alternative energy supply systems (Bándi, 1993, p.30). A decree of the Minister of Welfare authorizes smog alerts and stipulates the pollutant levels that will trigger such alerts (Minister of Welfare 5/1990.(XII.6)NM.rend. on Air Pollution Limits and Monitoring). Although formal authority in the hands of the inspectorates is rather significant, the more serious forms of its use are rarities. One close observer of the implementation process indicates no knowledge of any serious enforcement action, culminating in the prohibition of any industrial action, exercised in recent years.

The standards are both territorial and technological. Territorial standards, which are based on ambient air quality, are organized to deal separately with three types of sources: point, building, and surface. Technological standards are 'set based upon available technology and the quantity of the final product' (Bándi, 1993, p.30).

There are also three types of fines possible in this system. One is set via emission standards for the various kinds of sources. Emitters are required to self-report. The size of the fine is determined by the degree to which emissions exceed standards and the toxicity or hazardous extent of the pollutant, on a four-point scale. Fines are levied quarterly. Inspectorates have discretion to adjust standards to particular needs or situations. The shift in authority for setting individual emissions limits to the regional inspectorates was made in 1989 (Bándi, 1993, pp.30, 48; Government Decree 49/1989 (VI.5.)MT.rend.).

The regional inspectorates report to the Chief Inspector, who in turn reports to the Permanent Secretary of the Ministry for Environment and Regional Policy. The Chief Inspector is appointed by the Minister but is independent in operational terms from direct Ministerial control. Such feedback as is received at the level of the Ministry, in the Air Protection and Noise Control Department, typically travels through highly bureaucratic channels, and there is little direct connection between the enforcement staff operating from the inspectorate organization and those in the Ministry. These institutional features have impacts on implementation, as a later chapter of this study makes clear.

The regional environmental inspectorates have been limited by additional constraints in practice, including the need to rely on obsolete equipment in many regions. (International assistance, particularly through the European Union's PHARE program, is helping to remedy this problem. See United Kingdom, 1995, p.16.) Meanwhile, another feature of the current transition affecting the role of the inspectorates in practice is that, according to some analysts, their legal form is such that they can perform contract work in addition to their centrally-established responsibilities. Certain of these contractual efforts may place them into a conflict of interest with their central roles as enforcers of national policy. (For example, inspectorates have been able to contract with a local factory to monitor their effluent.) The financial pressures under which the inspectorates are placed contribute to this tendency and, some say, undermine their independence in practice.

To ease the information and monitoring tasks somewhat, the regulatory targets (permit holders) are required to self-report on emissions. However, until 1989 the report was only required to be an annual event; since then, quarterly reports have been stipulated. Nevertheless, and particularly during the process of institutional differentiation that has developed since the political changes, the inspectorates have faced a large and growing task in overseeing the self-reporting process and ensuring the accuracy of information and compliance with the law.

By relatively recent decree of the Minister of Transport, Telecommunication and Water Management, vehicles are now required to be tested for carbon monoxide emissions annually and are then issued a 'green card' as a permit for driving in traffic (18/1991.(XII.18.) KHVM.rend.).

Fines for air pollution violations have been one of the main sources of revenue for the government's Central Environmental Fund, although in recent years additional funding sources for the Fund, including bilateral and multilateral assistance from abroad, have reduced the proportion contributed by air pollution fines, and also by fines in general. In 1988, 384 million HUF came from air pollution fines; the total amount of all fines collected for the Fund was 495 million HUF; and the total income for the Fund in that year from all sources was 845 million HUF. The estimated 1993 amounts are 290 million HUF, 550 million HUF, and 3046 million HUF, respectively (Bándi, 1993, pp.62-63; see also Regional Environmental Center 1994a).

The decline in fines for air pollution has come primarily from better compliance by the regulated sources. The schedule of penalties has remained constant over the period. In 1992 a fee imposed on gasoline began to increase the income stream into the Fund significantly. Overall, the Fund receives revenue from fines, user fees, product charges, and international aid.

The Fund itself has been used to support environmentally-related expenditures. This financial mechanism thus earmarks certain revenues for environmental purposes. A regulation approved in 1992 broadened the authorized uses for the Fund beyond so-called end-of-the-pipe investments to include preventive efforts, and the possible financing instruments have also been broadened (Bándi, 1993, p.63; Regional Environmental Center 1994a).

## Additional Policies

Other policies with a rather direct impact on air pollution and its results could also be considered here. The most important additional policy sectors with direct bearing on acidification are energy and transport (see for instance Öko Rt 1992). As mentioned earlier, Hungary has embarked on efforts regarding transportation that are likely to exacerbate the air pollution problems from this sector in the future. Developments in the energy sector are also highly significant. One in particular is singled out for attention here: energy pricing.

During the state-socialist period, energy prices were set substantially below market rates. Energy was subsidized, and energy resources were imported easily from the Soviet Union. Since the political changes, however, the government has allowed increases in energy prices faster than the rate of inflation. These changes have been cause for considerable popular distress, and the government has been understandably reluctant to allow a full rise in prices to international market rates. The domestic pain has political repercussions. Energy prices are expected to continue to rise to market levels, nonetheless. To the extent that they do so, incentives have been altered to encourage cleaner, more efficient production processes and household practices.

Additional consequential government decisions that have had important impacts on acidification during the period of study are introduced later, in Chapters 9 and 10, where Hungarian responses to the CLRTAP protocols are analyzed. These have been most striking in the field of energy production, where shifts in earlier years to nuclear production and natural gas achieved large reductions in emissions. As explained below, these actions were products of governmental choice during the era of one-party governance, but they were undertaken for reasons apart from their environmental consequences.

Some decisions by the central government since the political changes have focussed primarily on environmental improvement as the motivating influence. On the question of pollution caused by vehicles, in particular, one of the few incentive-based environmental initiatives has been enacted. (On the limited array of incentive-based environmental instruments adopted thus far by the government, see Regional Environmental Center 1994d.) Starting in 1990, the excise tax began to be levied more heavily on leaded gasoline. The differential has since increased. Similarly, there are now (since 1992) certain tax benefits to individuals who use vehicles with catalytic converters, especially for smaller engines, thus offering modest official encouragement to those contemplating a shift. So there are some limited incentive-based instruments in this portion of the air pollution field (see Hungarian Commission on Sustainable Development, 1994b, p.17).

**National Picture: The Summary View**

In sum, the system for controlling air quality in Hungary is complex and difficult to administer. The standards are formally restrictive, but enforcement and ease of operation provide loopholes. Nor has this situation been eased in the transitional period following the political changes of 1989. Many new laws have been passed, though these are largely uncoordinated with each other. Changes have been made in the rules for controlling air pollution, but these changes do not get to the heart of the enforcement and coordination questions. A comprehensive recent review of the array of policy strategies and instruments used by several countries to seek achievement of emissions reductions for the CLRTAP identifies very few that have been successfully adopted and applied in Hungary (Economic Commission for Europe 1995). And the budgetary limitations remain severe in any case. The comprehensive new environmental policy has been accepted by the Parliament. However, an overall new set of programs, with implementation arrangements and realistic budgets, has not been enacted. The framework outlined above, therefore, can be considered as the backdrop within which acidification issues have been approached and can be expected to be dealt with for the immediate future.

**Local Government Involvement**

Prior to the political changes, independent local governments were not important to consider as autonomous public decision makers. With some exceptions, local councils were largely subservient to the central party-controlled system. The transition to a new political regime has altered this fact, with potentially-important repercussions.

The Fundamental Law on Local Self-Government charges the new local units with a range of significant responsibilities for environmental matters and also provides a set of powers formally available for their use. On the subject of air pollution, municipalities in principle have the primary responsibility for air pollution within their jurisdictions. They can also determine the amount of penalties assessed against violators.

In practice, however, local self governments have had to contend with contradictory regulations established by the national level, and they are overburdened with duties in a daunting array of policy sectors. Furthermore, except in rare cases they are not well equipped in terms of expertise or financial resources to assume a significant role on environmental matters. Nor have they been provided information on transboundary aspects of air pollution.

What is more, legal experts have seen the formal situation regarding localities' powers to control air quality within their jurisdictions as unclear. (Some very recent efforts at clarification may prove helpful at ameliorating this problem somewhat.) Their authorization to control air pollution has been based on the entitlement to regulate the services provided within their territory. Zoning regulations and permits for construction are the primary tools through which local self-governments can act to influence air quality. To complicate matters further, permitting is a function that formally requires the localities to act as agents for the central authorities, rather than in their own local interests. Yet much of the functioning of the new local units is designed (via the Fundamental Law on Local Self-Government) to aim at local control. The law now allows by implication the possibility that local units can impose stricter standards than those established nationally, but this possibility has not been explicitly outlined.

The issue has not been litigated, but should this possibility develop the Constitutional Court would be the ultimate arbiter. The lack of clarity during the several years since the establishment of independent local governments discouraged environmental activism even in those jurisdictions otherwise able to become active with respect to local air pollution problems. And the matter has not been aided by the tendency of some in the regional environmental inspectorates to seek to retain primary control over air quality rather than ceding some to the localities. Friction between localities and regional inspectorates is not inevitable, but it has developed often enough to limit the effectiveness of localities in addressing their air quality problems.

Some local governments, especially the larger ones, have contracted with consulting firms to assist them in assessing their environmental problems, and this activity appears to be on the upswing. Indeed, in 1993 29.8 percent of environmental expenditures of all sorts in Hungary were undertaken by municipalities - substantially more than the total expenditures by the Central Environmental Protection Fund and private business combined (Regional Environmental Center, 1994a, p.45).

However, the majority of local units have done relatively little thus far about environmental matters in general or air pollution in particular.

On the whole, then, national policy provides a formally imposing air quality control system which is seriously limited in practice by complex regulations, limited enforcement possibilities, relatively infrequent use of incentive-based measures, and budget constraints. Policies in sectors related to air quality but largely devoted to other issues also play a substantial role in the ultimate determination of pollution levels. Certain elements of the national scheme may develop into important and positive influences on air quality (for instance, the Central Environmental Fund), but during the period of investigation here these have been of relatively limited import. Likewise, the local governments now operating throughout the country may eventually become key centers of decision making for air quality, but legal and institutional impediments have circumscribed their role in the years since the political changes.

To understand more about the realities of environmental policy execution in Hungary during the last couple of decades, particularly for air quality, it is helpful to focus carefully on the institutional apparatus and procedures for administering programs and enforcing broad initiatives. The next chapter turns to this task.

# 7 Organization, administration, and implementation

While the general approach taken by the Hungarian government to issues of air pollution has remained relatively constant for a number of years, indeed even through the dramatic political and economic changes summarized earlier in Chapter 2, the actual execution of the policy and the institutional framework through which the policy is carried out have undergone substantial change. Both topics are important for the period covered in this investigation.

One dimension to consider has to do with the pattern of action under state socialism. A shortage of resources plagued the system for executing air pollution regulation under the earlier political apparatus, as it did during the first several years of the newer system. Perhaps even more important, however, was the politicization of regulation. State-central decision making provided a certain coherence and energy to policy pronouncements, but state firms and groups favored in the political apparatus were often able to violate the law with impunity. Regulations exempted certain industries that were deemed particularly important. And the court system was unavailable for judicial review of any of these decisions. The general public was effectively shut out of the process of policy making and enforcement (Bochniarz, 1992, p.13).

The political changes of 1989 have removed several of these obstacles: depoliticization of enforcement is taking place, privatization in any case alters the relationship between regulator and regulated, judicial review is now a standard part of the legal system, and broader participation by those outside government is gradually emerging. However, the complexities and problems of policy implementation have not disappeared.

## Implementation Consequences of Policy Design

Some of the most important issues affecting the implementation of air pollution regulation have to do with policy design. As suggested by the characterization of air pollution control measures given above, the basic approach taken in the law and regulations, including the complexity of the regulatory categories, tends toward obscurity and internal contradiction. Furthermore, the system of fines has served little as a deterrent. The total of fines collected amounts to only a small fraction of the environmental damage done by the pollutants, and firms have largely taken the system of fines into account as a part of their cost of doing business. Indeed, in real terms the fines have declined dramatically since the political changes of 1989, since the high rate of inflation (usually approximately 20 to 25 percent during this period) has eroded what little penalty-derived incentive firms might have had to comply with standards. There is little evidence that the fines have reduced air pollution in Hungary.

The need for resources continues as well, of course, and enforcers in the developing market economy are now facing an increased number of regulatory targets at the same time that enforcement budgets have been trimmed to deal with the current economic problems (see below).

In general, then, the strategy adopted for air pollution control during the period examined here has imposed significant burdens on implementers, and these have remained, for the most part, since the political and economic transition began.

## The Administrative System: Ministry for Environment and . . .

Beyond these points, the administrative system for policy implementation has also contributed to the complexity of the process. The institutional structure for carrying out air pollution regulation, indeed environmental policy in general, has been altered several times but remains subject to criticism from observers.

The structure for the administration of environmental policy in Hungary is focussed in the first instance on the Ministry for the Environment and Regional Policy, which in turn is organized into a number of specialized units. The current formal structure of this Ministry is the latest of several reorganizations since the late 1980s, and further restructuring has been considered during the tenure of the current Minister as well.

The Ministry, for instance, was reorganized in 1987 to place environmental issues and water management in the same administrative home. However, this arrangement lasted only until 1990, when water management duties were removed from the agency and placed with the ministry for transport and telecommunications. Most analysts believed at the time that the water supply interests dominated decisions having to do with water, at the expense of the protection of water quality. The reorganization was supposed to create a unit that would focus exclusively on protecting the environment and not developing or exploiting it.

In 1990 additional restructuring resulted nevertheless in the distribution of some environmental duties among numerous other units, including the offices discussed below for air quality. More modest structural changes have taken place a couple of times since 1990. The formal arrangement of the Ministry is depicted in Figure 7.1, which shows the organization as it existed in early 1996. Yet the Ministry has not enjoyed a strong reputation even in its revised form. For instance, 'the environmental committee of the Parliament was on strained terms with the Ministry for the past four years. Reports on the activities of the MoE to the committee were qualified as unsatisfactory' (Regional Environmental Center, 1994b, p.A13). The new government will undoubtedly undertake further changes, but their direction and shape is not yet clear, even to insiders.

This unit performs many of the national government's regulatory (and other) environmental duties, particularly in terms of establishing rules and broad policy guidance. Yet it also houses additional functions, certain of which seem to critics to be tangential to the environmental portfolio. Critics have observed, furthermore, that throughout the successive incarnations of the environmental presence within Hungarian administrative apparatus over the years, the structure has always been an 'Environment *and* . . .' arrangement, with the environmental issues always included (and, some would say, in competition) with such other goals as water supply or regional policy. While more encompassing administrative structures can assist in encouraging thorough decision making, they can also submerge environmental considerations in favor of others. The latter development is the concern of many observers, and the environment has been a persistently weak and somewhat grudging portfolio both before and after the political changes.

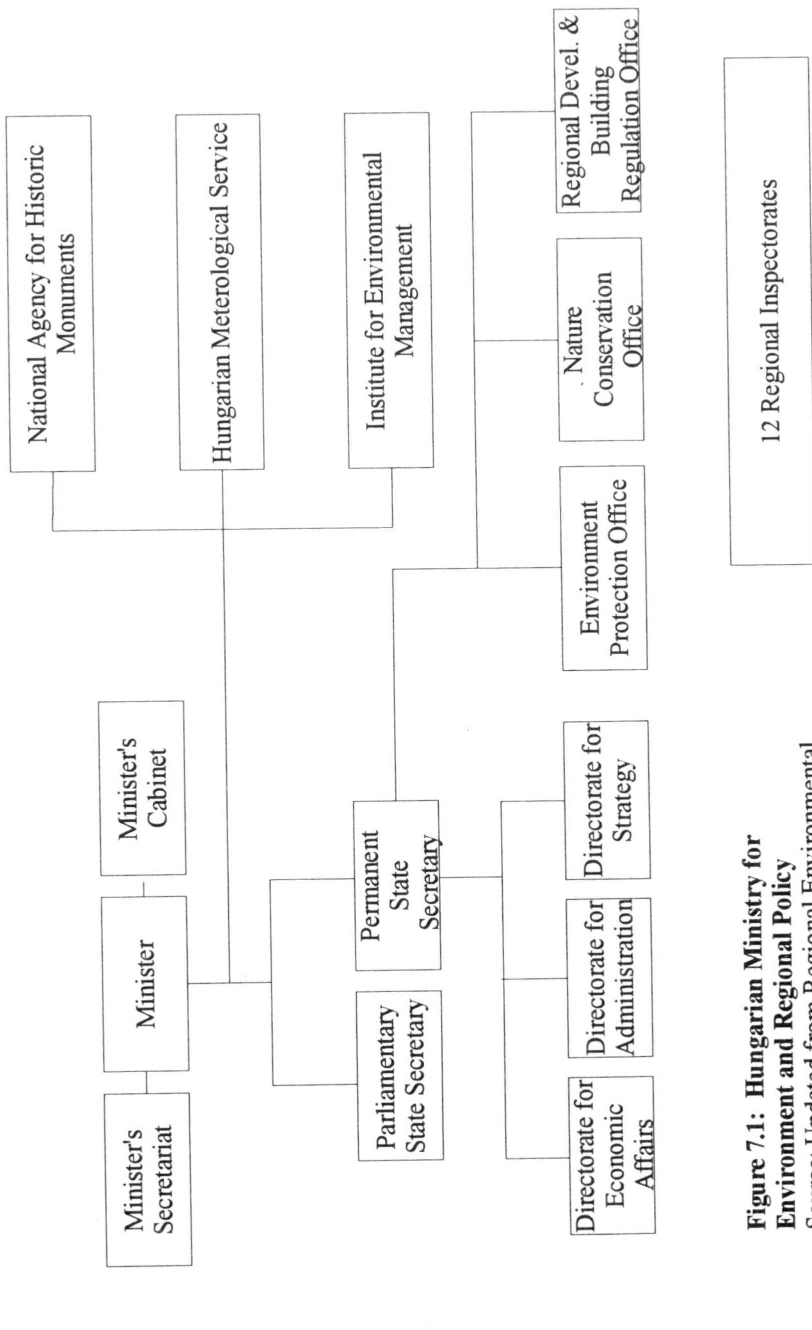


**Figure 7.1: Hungarian Ministry for Environment and Regional Policy**
Source: Updated from Regional Environmental Center 1995: 69

**Regional Environmental Inspectorates**

The heart of the field presence for enforcing environmental regulation, including those aimed at air quality, is in the Regional Environmental Inspectorates, of which there are 12. These units have remained in place through the recent reorganizations, and they operate in practice with considerable autonomy from the central ministry in Budapest, although the chief inspectorate, to which the regions report, is appointed by the Minister. The regional units serve as the field enforcement institution of the central government.

The regional inspectorates are currently in a position which provides operational evidence of the limitations in institutional capacity which are discussed more carefully below.

> Currently, [the regional inspectorates] have little incentive or motivation. Their financial basis should be more substantial. At this moment, unfortunately, inspectorates work partly as government authorities and partly as private consulting firms because they are only partially supported from the state budget, creating a strange situation. [Regional Environmental Center, 1994c, p.40]

One way in which this dual role manifests itself is through the process of environmental impact assessment. The inspectorates are charged with reviewing assessments before, for instance, construction of a new plant is begun. However, sometimes the inspectorate itself has been hired to conduct the review, thus involving itself in a conflict of interest.

The inspectorates are generally in untenable situations regarding their capacity to conduct their official duties. They received the short end of the resources 'stick' when they were separated from the water management directorates at the time of the 1987 ministerial reorganization (Hanf and Roijen 1994; Regional Environmental Center, 1994b, p.A13). And one of the most challenging problems they face currently is how to serve effectively as the primary filter, facilitator, and source of technical assistance for local and regional requests submitted to the Central Environmental Fund. Being strapped for resources, they sometimes do not operate effectively in connecting ideas for initiatives in the field with the priorities and trends as seen from the capital.

**Additional Units and the Division of Labor**

Beyond this institutional insulation of regional decision making from the central national policy unit, furthermore, another important pattern of differentiation separates execution of national policy into a number of different functional authorities, depending on the issue under consideration.

The arrangements for air pollution reflect this complex institutional patterning. The regional environmental authorities control the emission of pollutants, and they are charged with establishing emissions standards. However, ambient air quality standards inside settlements are set elsewhere in government: the public health authorities of the Institute of Public Health, located within the Ministry of Welfare. Because of the organizational locus for this task, the ambient standards are set with human health concerns uppermost.

Meanwhile, mobile sources of emissions are the province of the Ministry of Transport, Telecommunication and Water Management. Supervision of the vehicle inspection program is in the hands of the Public Transport Supervision Offices, and private firms can also undertake inspections if the firms have been licensed by the Offices. Violations are to be enforced by the Offices or the police.

Furthermore, the National Institute of Public Health is responsible for measuring background air pollution away from settlements.

> Operational since 1974, the National Imission [*sic*] Measurement Network has been measuring $SO_2$ and $NO_2$ contents of the air sampling sites and quantities of sedimenting dust at 680 sites. . . . In addition to the existing off-line system, a measuring network has been established which is capable of measuring meteorological parameters of air, its CO, $SO_2$ and $NO_2$, hydrocarbons, ozone and suspended dust and can transmit measurement results integrated at 30 minute intervals to the local data center (the subprogram was completed in 1993). [Hungarian Commission on Sustainable Development, 1994b, p.26]

The Hungarian Meteorological Service provides support for air quality objectives as well. And the nation's energy inspectorates have also been involved in furnishing technical assistance in this sector, which has been so important to emissions contributing to acidification. However, the shift to privatization has made it more difficult for these experts to advise firms, and firms themselves have been reluctant to employ energy experts during the recent economic difficulties; some have eliminated positions formerly devoted to this specialty. Currently, the energy inspectorates are slated for elimination via a government reorganization.

These several kinds of units have informed each other of pending new regulations by circulating drafts, but little regular and predictable coordination occurs following the setting of these central policies. No standing committees (below the level of the Cabinet) mediate regularly among them. In some locales, informal coordination among field officers of the units allows for a more integrated approach. But this result, highly dependent on the individuals involved within a given region, varies greatly in practice from one part of the country to another. The informal ties were also, reportedly, functioning better several years ago, before the political changes disturbed the stable patterns of patronage and personalistic coalitions that had marked the previous regime.

A summary depiction of many of the relevant monitoring units and channels in the Hungarian governmental structure is shown in Figure 7.2.

## The Privatization Process

In practice, as well, other ministries and agencies of the central government are involved on a day-to-day basis in issues affecting environmental decisions, including air pollution in particular. An important instance during much of the period since the political changes has been the State Property Agency (SPA), a governmental unit responsible for - among other duties - selling state assets into private hands through the privatization initiatives. The SPA was charged with facilitating the sale of several categories of formerly state-run firms, many of which contribute through their emissions to the acidification problem.

The privatization process itself has been complicated by the question of legal responsibility for environmental damage done by the firms, thus the environmental ministry and the SPA have had to interact during and sometimes after the property transfer process. Currently, as well, regulations require a sort of environmental review (not really a comprehensive audit) prior to the sale of a firm, with the product submitted to the regional inspectorate and ultimately the Chief Inspectorate and the

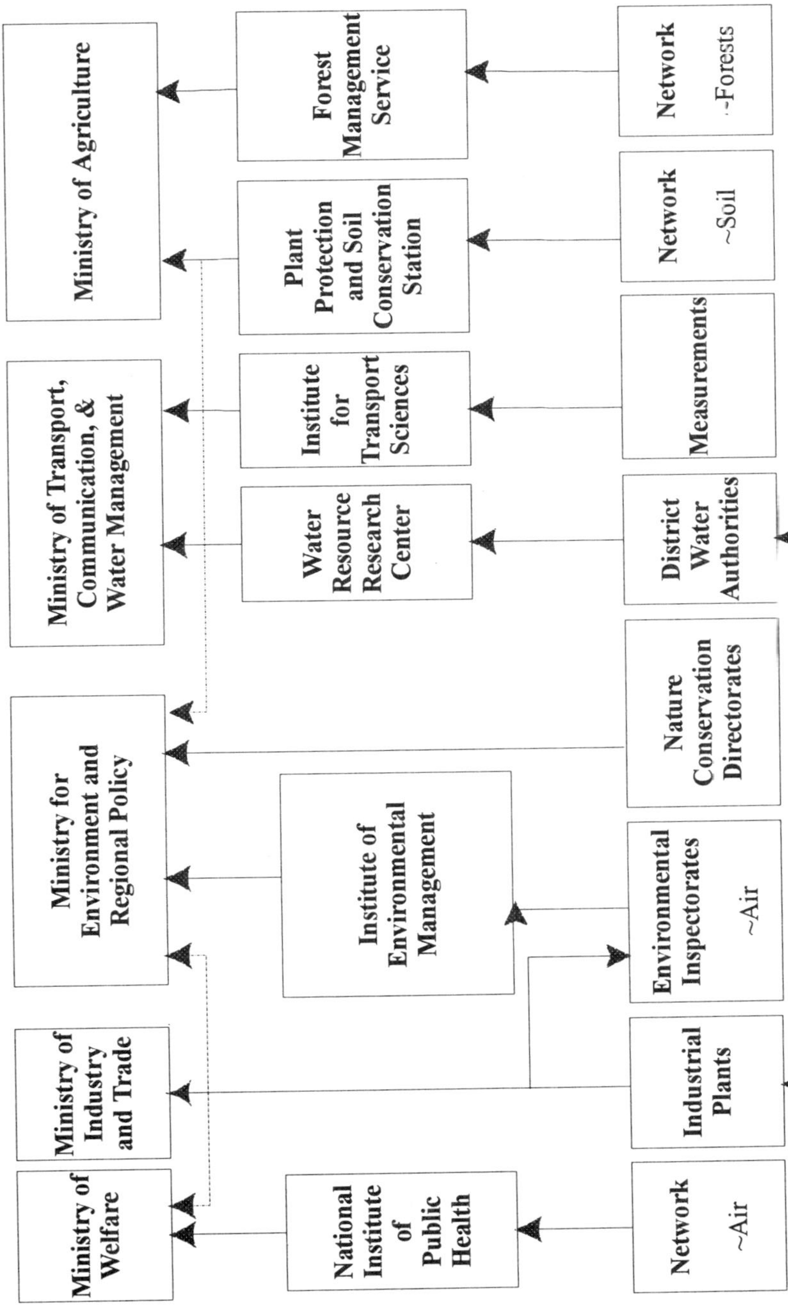


**Figure 7.2: Hungarian Environmental Monitoring Network**
Source: Updated from Regional Environmental Center 1995:70

environmental ministry. In cases in which environmental damage is discovered, the price of the asset might be reduced. Similar procedures are being used for bankrupt firms undergoing liquidation. Proposals have recently been drafted that would increase the likelihood that liability for environmental damage could ultimately be assessed, in a legal sense. Yet the government's policy has been somewhat vague. And this circumstance itself has hindered certain privatization efforts, particularly those cases in which foreign investors - who are reported to be more cautious on the issue of environmental liability than are domestic ones - have considered involvement but have been skittish under the relatively unsettled legal conditions.

Another privatization-related issue connected with the environment and air pollution has been what to do with those (portions of) companies that do not attract investor interest because of their precarious profitability. How should the environmental damages caused by these firms - in the past and also likely in the future - be treated? A strict requirement of compliance with even current standards, not to mention newer ones already drafted and likely to be enacted in the future (see below on European Union requirements in particular), would drive at least some such firms into bankruptcy, thus further complicating the liability question and exacerbating the economic troubles facing the country. Should these be granted a somewhat longer time to come into compliance? There has been no official policy on this tricky set of tradeoffs.

And a basic problem with the privatization issue is that in many cases little or no information is actually available on pollutants remaining at a site, nor on materials that had been used in production processes during previous years. And information on longer-term effects has generally not been taken into account in the privatization process and is unavailable to municipalities and others.

A further complication of the environmental challenge under the current process of property shifts occurs when industrial firms are shifted into the hands of local self-governments. For one thing, the environmental problems in these cases may be serious but the locals are often in poor position to fund cleanups or shifts in technology to ameliorate recurring problems like air emissions. Some observers believe that the most appropriate policy instrument here would be the creation of a fund to address such local environmental problems. Currently, nonetheless, the Central Environmental Protection Fund does support a number of municipal projects aimed at environmental improvement, including air quality efforts.

Local governments are also initiating new businesses, some of which have the potential to generate significant quantities of pollution. An increase in municipally-owned gasoline and service stations is a case in point. Some of these units of government may be caught between the conflicting desires to use such businesses as centers of revenue for local purposes and the policy goal of limiting traffic and emissions in their jurisdictions.

In addition, central government approvals for transfers of ownership have resulted in a multiplication of legally-responsible targets for emissions standards' enforcement, thus complicating the challenge of implementing regulatory policy under the currently-strained resources and institutional capacity of the Ministry for Environment and Regional Policy and the inspectorates. Nevertheless, the authorities hope that ultimately privatization will produce improvements in environmental conditions, since private firms are expected to organize production processes more efficiently than the state industries had done - thus for instance trimming energy use and operating more environmentally friendly technologies. (Note the earlier coverage of hypothesized relationships between these property shifts and levels of air quality.) The issue is treated again later in the analysis of efforts to achieve national goals under the CLRTAP.

Hungarian policy on the handling of environmental damages and liability for companies facing liquidation procedures has evolved during the post-state socialist period. A law on the subject of the rapidly escalating liquidations came into force in 1992, and an amendment the following year was aimed at dealing with environmental impacts. Cleanup costs, plus the expenses of bringing facilities into compliance with standards for ongoing operations, were to be included as part of the overall total of liquidation expenses. Thus these were to be dealt with before, for instance, banks would receive any assets. If trustees during the liquidation process were to fail to take care of the environmental issues themselves during a two-year transition period, the liabilities would be transferred to the new owner(s). However, the process and ultimate placement of liability have been somewhat confusing to those in the system. Additional governmental decrees have been prepared with the object of clarifying these complications.

A new fund established in 1996 reserves a portion of the income from the privatization effort for helping to cover liquidation costs and also expenses related to addressing environmental damages as uncovered in the property transfer process.

Further, the practical job of actually enforcing the existing regulations, which falls primarily to the regional environmental inspectorates, has been complicated not only by the difficult economic circumstances but also by the sheer increase in numbers of regulatory targets that have followed from both the new 'greenfield' industrial firms that have been created and the dismemberment of large state-owned plants and firms into numerous smaller units - some sold into private hands, others (often those in the worst economic and/or environmental condition) remaining with the state, still others having been moved into private hands (again, see the earlier coverage of the hypothesized relationships).

The law stipulates that when liquidation begins, the regional inspectorate must be informed. The firm involved has to file a report on environmental conditions. Documents of some sort are typically submitted, but these are sometimes not satisfactory. And the inspectorates themselves are often overloaded, their extensive responsibilities precluding careful independent investigations to verify information and gather additional data. Internal calculations of the Ministry for Environment and Regional Policy on the budget necessary to bring the inspectorate staffing up to levels necessary to execute responsibilities under the liquidation process show that an additional HUF 40 to 50 million per annum would be needed. This figure can be interpreted as approximately another one or two persons per regional office. However, the central government's austerity programs, affecting many policy sectors as well as the environment, have starkly limited the options for the near term at least.

Another complication of the process in practice is that the environmental ministry, which has worked to become involved in the property transfer decisions, has become seen as a bottleneck itself. 'Too few people chasing too many problems', is the way one insider describes the situation during this transitional period. The property transfer bureaucracy in the central government, being charged with completing the transactions and thus satisfying the investors, has found itself hindered by the need to deal with these 'additional' considerations. The two parts of the national government (the Ministry for Environment and Regional Policy and, for most of the transitional period, the State Property Agency) have made attempts to work together, but the relationship has been persistently complicated. One environmental expert was eventually employed by the SPA, thanks to a program of bilateral assistance from the United States. And liaison efforts continue between the property transfer and environmental agencies.

So too do efforts to develop approaches that encompass the local governments' new roles in liquidations. A bill considered recently would provide localities with a fraction of future emissions charges (drawn from property within their territorial jurisdictions) to support local environmental efforts. Whether this instrument might provide an incentive for localities to encourage further end-of-the-smokestack approaches rather than emission reduction and compliance-oriented strategies (by providing financial payoffs for violations within municipalities) has not been considered within the Ministry. Indeed, this proposal provides further evidence of the persistence of the earlier Hungarian approach to emissions down to the present day.

For all of these issues connected with the massive property transfers underway, the tasks are likely to become even more challenging in the next period. The newer coalition government has accelerated the privatization process since 1994, indeed the increased pace provided a substantial segment of revenue for the government during the preceding year. The pace of change has thus increased as well. And earlier experience with rapid property shifts (for instance, with protected forest lands sold into private hands, only to have this transfer reversed by the national Constitutional Court because of previously-established legal protections - with great confusion as a result) suggests that the environmental questions that need to be addressed will require care and resources for appropriate resolution.

## Status and Influence of the Ministry for Environment

When dealing with environmental policy issues in this larger institutional and political matrix, the environmental ministry has often been not especially influential. The quoted excerpts above (Chapter 5) evaluated the influence of this ministry as limited, and this conclusion is reinforced in discussions with knowledgeable observers. Several actors involved in the field in Hungary report the Ministry's role as 'quite weak', or synonymous terms, particularly by comparison with ministries in certain other policy sectors. The reasons seem to have to do in significant part with the economic and social crises afflicting the nation - and the attendant inflation, unemployment, diminution of social welfare protections, and other highly salient issues.

A summary of the state of affairs is provided in a recent report from the Regional Environmental Center:

> The MoE [Ministry of Environment] does not have enough leverage to negotiate with other ministries for the incorporation of environmental considerations into the policies of other sectors such as energy, finance or economic development. According to some experts the MoE needs more highly trained environmental specialists, often right to the top level. Others criticized it as being too political. The budget allocation for the ministry provides an insight into the priority given to environmental protection by the government (0.6% of GNP is allocated for environmental protection whereas environmental damage is estimated at 3-10% of the GNP). [Regional Environmental Center, 1994c, p.40]

The foregoing comments point in detail to a general pattern with regard to acidification policy in Hungary, a pattern that requires broader comment and interpretation here. Discussing the issue means introducing here some topics related to the CLRTAP coverage in Chapters 9 through 11.

Although many aspects of Hungarian policy on air pollution issues can be criticized, a major difficulty stems not from lack of reasonable standards and boldly-stated objectives but, rather, from lack of implementation. This point bears directly on the international regimes for acidification, and the reciprocal influences of these regimes and the actions of the government of Hungary.

Especially in recent years, as documented below in Chapter 9, there has been no real difficulty in getting Hungary to agree to participate actively in the negotiations - culminating in adoption and ratification - of the protocols on LRTAP. The renegotiated $SO_2$ agreement of 1994, for instance, was accepted by the Hungarian government with relative equanimity. (The regulations of the European Union regarding large combustion plants and vehicle emissions, now relevant because of Hungary's status as associate member and commitment to achieve full-member status as soon as possible, also have to be dealt with as a matter of government policy. Proposals have been developed for revising Hungarian law to achieve conformance with EU regulations on the environment more generally.)

One reason, undoubtedly, was a strategic calculation that the emission reductions required under the revised protocol, although substantial, have a high probability of being achieved without additional painful governmental steps (see below). The expectations of the large energy firms (backed by the incentives they now face) to move away from heavily polluting fuels and toward cleaner technologies suggest reasons for optimism. Indeed, one manager of air pollution regulation in the Hungarian government indicates that the government sees the $SO_2$ problem, at least, as largely 'solved already' - even if the solution is not yet a complete *fait accompli* in the emissions data.

Another reason is the government's agreement with the goals of the international protocols of LRTAP and its willingness to commit itself to targets and principles. This willingness seems exhibited in the actions of the government in participating in the current $NO_x$ renegotiations.

Whether international negotiations have real, independent impacts on emissions in Hungary is more difficult to determine (see the coverage in Chapter 9 and interpretive comments in Chapter 12). Partially the question is complicated by the subtleties of methods and measurement. In addition, governmental commitments entered into in good faith are likely to face some constraints stemming from a lack of implementation capacity in the field of environmental policy in general.

For reasons just explained, the government may be in position to achieve 'success' in $SO_2$ implementation as a product of other causes not directly rooted in national policy on air pollution. But the broader picture shows a government that lacks significant implementation capacity in the field of environmental policy in general.

The reasons have partly to do with what Underdal has called the 'vertical disintegration of policy' as it moves from general national commitments to the specific demands (costs) placed on social sectors relative to the benefits they can expect to reap directly (Underdal, 1979, p.7; see also Hanf and Underdal 1996). This disintegration, or perhaps benign neglect, is most apparent when environmental questions face economic ones in the difficult transitional period. Here the generally accepted environmental goals lose out in the competition.

Another aspect of vertical integration, unanalyzed but probably assumed by Underdal, is support from above. Support from a parliamentary majority is obviously necessary for coherent policy making and subsequent execution. In the Hungarian context, however, support from above is necessarily limited. Even if one leaves aside the competition with pressing issues like the economy, parliamentary majorities in this transitional period are unlikely to be strongly supportive of environmental initiatives. The two governments of the post-socialist era have been coalitions drawing from differing perspectives; but no green party commands parliamentary strength, let alone a place in either governing coalition. Informal support from shifting clusters of legislators is insufficient to support firm direction or a set of energetic initiatives.

In addition to the vertical dimension is a more 'horizontal' one: the ability of central government to involve important sectors of the public in policy discussions and, potentially, in mobilizing support for new initiatives. On this score different political systems develop these links in different fashions, so - for instance - pluralist and corporatist patterns of interest involvement can be markedly different, with real consequences for policy and practice. In Hungary and other formerly one-party systems, of course, the mobilization of public 'support' for official actions has left an aftertaste in the current transitional period. Those in the wider society with, for instance, deep environmental concerns have been wary and somewhat distrustful of government action; even on those occasions when involvement in real policy discussions has been possible, motives have been suspect. And government officials have had little experience with or interest in sharing influence and dealing in good faith with such 'outsiders' as NGOs. When combined with the lack of experience on the part of some NGOs and others, these inclinations on the part of government representatives have translated into relatively weak and unproductive horizontal connections with potential allies or sources of support in the broader society.

In the case of the environment, and for the Ministry for Environment and Regional Policy, the lack of capacity is manifest not only in the relatively limited resources, influence, and vertical linkages within government, but also in the very limited ties with, and networking opportunities among, NGOs and others who remain seriously concerned with environmental questions and might be able to help catalyze support for environmental action. The Ministry provides no regular contact with the public, no dissemination of the results of environmentally relevant research, and no regular public information on the spending of environmental funds or related matters.

But the vertical disintegration and 'horizontal' gaps as sketched here are not the only considerations. It is not merely that environmental institutions are less powerful and exhibit less capacity than other, competitive sectors and institutions, although that is indeed the case. The explanation lies also in a broader institutional incapacity. Even in key policy sectors like privatization, the institutions for execution have lacked needed capacity and thus have labored with relative ineffectuality (see O'Toole 1994; Rondinelli and Fellenz 1993). In its transitional period, Hungary has faced the handicap of having to develop institutional capacity generally, across policy sectors, levels of government, and in the broader society rather than merely the state. All these sectors and levels are short on expertise, lacking in needed budgetary resources, and often handicapped by a lack of pressure group support from the broader society.

The relative ineffectuality of environmentally focussed institutions, then, does matter. And explaining the implementation gap on acidification issues does require comparing the relative support for environmental vis à vis other institutions and interests in contemporary Hungary. But it is important to bear in mind that in Hungary and the transitional nations of Central and Eastern Europe, institutional capacity is in short supply more generally. Any analysis that treats acidification in Hungary, then, as merely an instance of environmental institutions losing in political competition would be misleading and incomplete. The broader challenges and institutional needs of the nation are also a piece of the explanation.

All this discussion should not suggest, however, that the institutional needs of the environmental sector are not important to the acidification issues as they manifest themselves in Hungary; they are.

But these comments point clearly to the larger socioeconomic and political complications of any effort to develop better policies for air pollution control in Hungary at the current time. And they also suggest the need for a more developed institutional capacity for implementing such policies. It is to this latter issue that the discussion now turns.

# 8 Institutional capacity and development

The foregoing review of Hungary's approach to air pollution issues and related challenges suggests a number of needs within the country if air quality is to be fully effective in the current context. Obviously, policies can be improved; and some observers suggest the importance of the government's increasing the priority it now places on environmental issues.

Still, the most well-designed policies are worthless in the face of an incapacity to execute them. And high levels of commitment on the part of a government matter little if the commitment cannot be converted into effective streams of action. Hungary's capacity to both implement international agreements on acidification and also influence their revision over time is thus dependent in part on the nation's institutional capacity.

Further, as the coverage above makes clear, that capacity has been tested in recent years (Agh 1993; Hesse 1993; Szabo 1993; Verebelyi 1993). Years of underattention during the period of state socialism, a fragile economic system during the transitional period, shortages of financial resources and staff in the relevant ministries, a lack of effective channels of coordination among the several official units acting to execute national policy, and limited horizontal connections with those interested in stronger and more effective environmental efforts have all meant that limitations on institutional capacity have constrained what has been possible during implementation.

## Limited and Stressed Capacity

In concrete terms, and aside from limited budgets, the agencies involved in implementation lack: sufficient highly trained experts, institutional infrastructure such as well-developed information systems, managerial skills, technical resources like equipment and monitoring facilities, experience with pollution-preventive policy tools, experience with incentive-based policy instruments (of which, practically speaking, there are virtually none in the environmental sector), a history and recent tradition of nonpartisan public service, a pattern of openness to the broader public for decision making and the provision of information, standard and efficient interagency channels of communication and coordination, intergovernmental mechanisms of information sharing and collaboration, and experience in overseeing practical program execution.

Additional stresses on institutional capacity can be noted. The political changes were accompanied by widely-supported decisions to develop a market economy in part through the privatization of a substantial portion of state assets. The several privatization programs, and the inevitable liquidations that followed for numerous unprofitable enterprises, placed additional strains on the developing institutions of the mixed economy as well as on the government that was seeking to stimulate economic success.

The political changes also brought increased challenges to the 3200 local self-governments, units now charged with substantial responsibilities in numerous policy fields but in need of considerable institutional development themselves. And the local units were further pressed by revenue shortfalls, as the national government began to trim the aid to localities from the center. Local governments, then, which might have taken some of the responsibility for environmental protection, have been already burdened, or overburdened, with other tasks.

Another aspect of institutional capacity has to do with the institutional presence and strength of nongovernmental organizations (NGOs) that are interested in and work on behalf of environmental issues. NGOs began to develop actively in the 1980s, as the discussion earlier regarding the Gabcikovo-Nagymaros Dam dispute indicated. In fact, political opposition more generally rallied under the environmentalists' banners (see Regional Environmental Center, 1994b, p.36).

NGOs have begun to participate in the public debates surrounding environmental issues in the post-state socialist setting, as well. For instance, they provided input in the extensive discussions surrounding the new environmental policy considered by Parliament during the last couple of years. But the general waning of salience of environmental issues in the public at large and among the political leadership affects their degree of influence, and the lack of regular access on the part of these organizations to government decision makers also limits their involvement, as explained in the preceding chapter.

Hundreds of NGOs operate actively now in Hungary (the Regional Environmental Center counts more than 200, and this may be an underestimation; see Regional Environmental Council, 1994a, p.A8), and some of these focus on air quality issues - either exclusively or as a priority item. Yet water issues, particularly the contamination of drinking water supplies, are seen as the highest priority in Hungary currently.

Even for those NGOs focusing on air quality issues, however, multiple subissues consume time and attention. The Clean Air Working Group, for instance, focuses its main efforts on improving air quality in Budapest. Attention on international matters is devoted to Hungary's compliance with the Montreal protocol on ozone, rather than on the issues arising directly from commitments under the CLRTAP and the regulations of the European Union. Similarly, the Energy Club is active on its subject but uninvolved with CLRTAP.

Therefore, the institutional capacity beyond the government for support of compliance with international obligations regarding acidification itself remains fairly limited, and its ability to assist in the decisions and implementation of acidification questions minimal thus far. Still, there are possibilities for institutional development and a leveraging of outside pressure that were simply not possible under the previous political regime.

The Hungarian private business community itself, collectively speaking, has recently begun to consider environmental questions as a part of its agenda. The Chamber of Commerce has created a position for an environmental officer to deal with such issues on behalf of business interests. Of course, this development is relatively recent, and the actions of the institutionalized business community in the environmental sector - whatever they may be - lie primarily in the future. (See the coverage later for analysis of links with individual firms.)

A further institutional limitation derives from the system for public data gathering and monitoring, which is weak, underfunded, and by some estimates deteriorating in the context of the transitional setting. On the other hand, private sources of data are beginning to improve. (And, critics charge, some regional inspectorates are taking public data 'on the market' by selling it to interested parties.) Some concern is voiced by environmental specialists in Hungary that the advantage the government has had in information and monitoring may be in the process of being lost, with eventual damage to the quality of environmental policy and protection.

A prime example of this deterioration is said to be the increasingly tenuous situation of the Institute for Environmental Management (see below), which was viewed as having performed its role with some distinction and quality in earlier years but is now threatened by budget cuts, staff exodus, and a lack of institutional definition and direction.

Nor is this kind of role - that of the institution specializing in quality expertise and analytical capacity on environmental matters - being filled by other candidate successor institutions. Several universities have begun to develop departments and programs around the specialty of environmental science, but public support for universities is being trimmed during this period of austerity. Student demand is consequently controlling funding - that is limited in any event - to an increasing extent.

And the independent research network of the Hungarian Academy of Sciences, in some respects a typically-Soviet creation, contained under the earlier political system some significant centers of quality work on environmental issues. The financial exigencies are pushing these elements into the realm of contract research, thus leaving little room for such essential environmental tasks as monitoring, methodologically-rigorous research, and other similar efforts.

In this context, then, the Hungarian government has had need for significant expansion of its own institutional capacity as a part of any concerted effort to address environmental issues.

Parliament, as indicated earlier, has not been a source of leadership on the environment. It is also not organizationally well-equipped to serve as a center of environmental awareness or review. The only forum in which environmental issues are regularly discussed at the legislature is in the relevant committees of the respective political parties.

With regard to air pollution, and also for other fields of environmental policy, governmental managers agree that coordination has not been enhanced by the formal bureaucratic structure. Indeed, what coordination does occur, especially across ministries, is accomplished through patterns of informal relations.

Part of the difficulty here is the legacy in Hungary of strongly-hierarchical institutional structure, exacerbated by official discouragement during the state socialist regime of informal connections - 'networking' - at operational levels. This experience, plus the relative lack of a strong, independent civil service culture within the agencies of government, inhibits the complex interconnections that would be needed for effective implementation on a regular basis.

The judiciary as an institution is also not ideally formulated for easing the execution of environmentally strong policy. In Hungary the law sets out general principles, but the judiciary is reluctant to elaborate these into the often-crucial details of individual cases. The process of litigation, then, is typically not an efficient route to environmental protection.

The government - both the first coalition and the current one - has made some efforts to expand institutional capacity in the field of environmental policy and its implementation. Among the efforts made have been the following: reorganization of the Environmental Ministry, establishment of an environmental impact assessment process, recent placement of an environmental specialist in the state property management bureaucracy to help with the privatization and liquidation processes, and the use of aid from abroad for additional institution-building efforts related to the environment. The result has been some expansion of capacity, but only to a limited extent. Hungary stands in need of additional capacity building in this sector. Each of these efforts can be discussed briefly.

## Recent Capacity-Building Efforts

Reorganization of the Ministry for Environment and Regional Policy has become a relatively frequent event in Hungary. The widespread expectation is that another structural revision will occur before long, particularly since most parties involved in environmental policy in Hungary profess unhappiness with the current arrangement, especially because of widespread criticism of its coordinative capacity. The Ministry has also made some efforts to invigorate the organization with new specialists, some of whom are drawn from the NGOs and scientific community within the country. Limited salaries, however, and business opportunities in the private portion of the mixed economy constrain this infusion of talent.

An environmental impact assessment (EIA) process is now required for significant new development projects. This process is meant to be a systematic review in advance of environmental aspects of new projects, including potential air quality impacts. In practice, however, the quality of these reviews is highly variable. There is no certification procedure for the firms that might be involved in the assessments, for instance, and the inspectorates and Ministry review only a sample of the more important EIAs produced. As a consequence, the EIA requirement does begin a process of institutionalized, proactive approach to environmental issues including air pollution, but thus far it has had limited impact.

Similarly, the placement of an environmental specialist in the bureaucracy focussed on privatization is a start at the development of institutional connections, and more complete balancing of issues within institutional contexts, that can aid the process of coordination and coherent policy execution in Hungary. Yet this initiative has involved only one position, and the basic national policy on how to treat environmental issues in the privatization (and liquidation) process has required extensive clarification. And even modest related efforts - like the creation through the Ministry of the Environment of a committee, including two environmental experts, to advise the privatization apparatus - can also be interpreted as signs of progress at only glacial speed. In this example, the process of committee creation took two years.

Finally, some coverage of aid from outside the country for institutional development can be included. Support from abroad for environmental improvement derives from several types of sources, such as the PHARE program of the European Union, bilateral programs of the OECD countries, German Coal Aid, the US Agency for International Development, and foreign NGOs (Regional Environmental Center, 1994c, p.42). As the figures below suggest, the majority of this assistance, including the majority of aid for combatting air pollution, derives from the EU.

The number of air pollution projects supported internationally (excluding aid from the World Bank) totalled 24 as of 1993, with the European Union leading in assistance (in terms of total numbers of projects) with ten. Bilateral aid was received from the Netherlands (five projects), Belgium (two), Denmark (two), and the US, Canada, Germany, and Austria (one apiece) (based on Bándi, 1993, p.143, which in turn is derived from a review of the files of the Environmental Ministry).

Contributions in ECUs for air pollution assistance in Hungary have also been led by the efforts of the European Union, which had contributed 13.775 million ECU as of 1993 for air pollution control efforts alone (and a total of more than 31 million ECU of environmental assistance overall). Several programs of bilateral support have also aided the Hungarian government's air pollution efforts. The total contributions from all these sources (excluding the World Bank) for air pollution programs was 17.119 million ECU in 1993 (calculated from Ministry files and reported in Bándi, 1993, p.145).

Among the efforts to improve air pollution control have been support from PHARE to provide equipment for the on-line measuring network, and aid from Japan for equipment for monitoring in the Sajo Valley, one of the most heavily air-polluted regions of the country (Hungarian Commission on Sustainable Development, 1994b, p.26).

PHARE in particular has come to be relied upon in increasing measure for some aspects of institution building within Hungary. Still, the potential of PHARE - even if all the program's investments were to be soundly conceived and executed - should not be overestimated. In a recent years this channel of assistance contributed only five percent of the nation's total environmental investment, even in these relatively lean years for domestic programs (see United Kingdom, 1995, p.11).

A significant portion of the assistance from abroad has been devoted to what might be termed, in general, institution building: support for personnel, aid for planning and analysis, the development of information systems, enhancements in technical capacity - via personnel, training, or purchases of technology -, and similar efforts. Some has also been devoted, clearly, to purposes directly connected with the international acidification regime. Norway, for instance, provided a measuring station within the framework of EMEP. Other countries have provided measuring equipment and related support.

How useful has this assistance been in the Hungarian context? A balanced assessment can be excerpted:

> Western countries have offered substantial support to Hungary during the transitional period. The most important aspects of this aid are in the form of knowledge transfer, help in institution building, and directly financing projects. The experiences with western assistance have not always been positive. Some consider the main goal of western consultants to be making money, which means that as little effort as possible is put into the work. Some 'assistance' projects are organized in such a way that consultants come, write a report and leave again. This is, in fact, only transportation of money from the left pocket to the right, i.e., Western assistance is

> sometimes merely self financing, which really does not help Hungary that much. It is noted that Hungary is also guilty of this practice. [Central and Eastern European] countries have not learned enough regarding the proper use of financial resources and they also lack experience to tackle environmental issues. Western programs, like PHARE, have very strict rules, making it difficult for countries to comply and receive assistance. [Regional Environment Center, 1994c, p.42]

PHARE, as one of the major programs of assistance, has not been immune from both negative and positive characterizations in this regard. Many observers consider it to be a helpful stimulant for the development of democratic institutions, but its cumbersome administration and the lengthy delays sometimes associated with approvals and oversight have been frustrating for many involved. Relations with NGOs have suffered as a consequence, as well (see United Kingdom, 1995, p.19).

The same analysis excerpted above offers sound, if not particularly innovative, advice on how to improve institution-building efforts financed in part from abroad:

> Hungary should make use of Western assistance and, therefore, develop better diplomacy by trying to combine the EU's interest with our own interest and then, for example, ask for more investments in alternative transport. International programs should be reformulated and the objectives should be adapted to the needs of the countries. Furthermore, western consultants should pay more attention to the local circumstances. [Regional Environmental Center, 1994c, p.42]

Subject to particular criticism have been many of the 'feasibility studies' supported by the aid available, particularly when the information to determine feasibility has been already available and the most urgent need has seemed, instead, to be support with the actual projects themselves. Considerable friction and disillusionment have been engendered by this pattern of assistance.

Regardless of one's assessment of the most appropriate course to take, it is clear from an assessment of current institutional conditions that Hungary's pattern for dealing domestically with issues of acidification is one of limited capacity. It is unlikely that radical changes will be seen in this picture in the near future.

For purposes of dealing with the international acidification regime exemplified by the CLRTAP, however, and particularly over the last couple of decades during both the current and the preceding domestic political and economic systems, how has the nation responded? The full answer to this question must consider Hungarian institutional capacity to formulate choices and put these into effect, so the issues raised in this chapter bear on the research questions of central interest. So too do the full array of contextual features outlined in the preceding parts of this inquiry. At this point the volume turns to the issues of central concern: the nature of Hungarian involvement in the LRTAP regime; the formulation of national positions on the main questions of international concern; the extent of influence by the international regime on domestic policy making, implementation, and outcomes; and dynamics over time, including changing national approaches in response to endogenous (policy-oriented learning, strengthening of and support from the international level) and exogenous (economic and political) forces.

The next three chapters address these topics.

# 9 The Hungarian approach to LRTAP

The involvement of Hungary in the international acidification regime that is centered on the Convention on Long Range Trans-boundary Air Pollution (CLRTAP) of the Economic Commission of Europe has stretched from the earliest agreements initiating the Convention down to the present day. Through both the older and newer domestic political systems, and despite dramatic social and economic changes spanning two decades, the nation has not wavered in its public commitments to the CLRTAP and its subsequently negotiated agreements, which have included international adoption of a standard monitoring system and a series of chemical-specific protocols stipulating quantitative targets for emission controls or reductions.

Nor has Hungarian involvement been restricted to the formal negotiating table. Representatives from the country have occupied leadership positions in LRTAP's working groups, cooperative international elements of the international regime; and - as this chapter and the one following document - the negotiated protocols have been treated seriously in domestic decision making and subsequent Hungarian efforts to participate in international efforts. Despite, and in part because of, the domestic disruptions in Hungarian social and economic life, the nation has thus far complied with all the stipulations of the CLRTAP and its protocols; and short-term projections indicate this pattern is unlikely to change in the near future.

The summary, and necessarily superficial, view must be, then, that Hungary's involvement in LRTAP signals a successful set of national-international linkages in the interests of environmental protection and transboundary cooperation. There is truth in this assertion. There are also, nonetheless, misleading elements to this generalization. Hungary's involvement in the acidification regime is both more complex and more interesting than any such straightforward conclusion.

The influence of an international regime like LRTAP on the perspectives and actions of a signatory country like Hungary can take place through a number of routes and via any combination of several causal paths. For instance, regimes like this one can place international pressure and attention on acidification objectives, help to stimulate trust and support among participants both domestically and internationally, encourage consideration for productive policy initiatives, provide information helpful in grappling with the problem, and render coordination among interdependent actors and nations easier and less tendentious. (For a review of some of the functions of such international regimes, see Helm and Sprinz 1995.)

The channels and possibilities are sufficiently complicated and numerous that it would be an impossible and largely unproductive task to seek here to separate out these multiple strands of influence for the Hungarian case. And, in a single-case analysis, especially one with a time horizon as short as that of LRTAP, there is no effective manner of conducting rigorous analysis to assess the relative importance of the several channels of possible influence from the international level.

The approach used here, therefore, is to assume that some combinations of these routes to influence domestic behavior on acidification are indeed facilitated or encouraged by LRTAP. (Some evidence for these influences along certain causal paths is presented where appropriate.) What follows, consequently, is an effort to trace the behavior and other evidence regarding Hungarian responses to the set of international commitments; and ultimately to assess whether and how much national decision making, implementation, and target-group behavior is affected by the regime commitments.

In this chapter and the two immediately following, the basic information regarding Hungary's involvement in the CLRTAP is presented. The presentation begins with an overview of Hungary's chronological involvement and set of formal commitments. Detailed material regarding the overall CLRTAP and its initiation, governance, and operations is omitted in favor of concentrating on the Hungarian features of the experience. The Hungarian government's general approach to LRTAP is sketched, and the most important institutional settings for acidification issues in the domestic setting, aside from the general institutions for governance and environmental policy described in earlier chapters, are sketched: the Interministerial Steering Group and the Institute for Environmental Management.[1]

This chapter then outlines the processes as they have developed in Hungary for deliberation, decision making, and implementation regarding international commitments through LRTAP. Interestingly, the main participants and processes for decision making have not changed dramatically from those in place under the one-party system; but some of the implementation challenges have escalated, particularly during the recent transitional period. Some details pertaining to the main chemical-specific protocols and their consideration in Hungary are presented in Chapter 10. And both Chapters 9 and 10 contain by implication information bearing on the set of research questions outlined in the Introduction and the additional hypothesized influences on acidification, as presented in Chapter 3.

## Hungary and the International Acidification Regime: An Introduction

As explained earlier in this volume, Hungary's involvement in LRTAP was originally stimulated in part by East-West political tensions, rather than simply by a straightforward environmental agenda. The section following this one explains, furthermore, that additional considerations influenced positions taken by Hungary early in the negotiations on a CLRTAP. First, however, it is helpful to sketch in outline form the formal commitments undertaken by Hungary during the course of the LRTAP effort.

Hungary was a signatory nation to the original LRTAP Convention negotiated in the late 1970s. Hungary signed on the international adoption date of 13 November 1979 and ratified the agreement on 22 September 1980. This basic commitment on the part of Hungary and other participating nations was to accept in principle the goal of reducing transboundary air pollution contributing to acidification and to work toward practical reduction steps in subsequent years. Hungary, along with other countries, agreed to keep other participating nations informed regarding national policies and programs related to air quality, provide emissions data, and participate in international research programs.

The next international commitment emanating from LRTAP was the establishment of the so-called EMEP system (Cooperative Program for Monitoring and Evaluation of the Long-Range Transmission of Air Pollution in Europe) for measuring transboundary flows of pollutants.

The EMEP protocol to the convention, viewed by many analysts as an essential element for subsequent emissions reductions (for instance, Levy 1993), was focussed on international adoption of an acceptable measuring system upon which quantitative targets could be based. The function of EMEP, whatever the system's innate strengths and weaknesses, was to establish a sufficient level of information and therefore the basis of at least minimal levels of trust in the international context, so that the acidification regime would not collapse into ineffectuality. The EMEP protocol was signed by Hungary on 27 March 1985 and ratified on 8 May 1985. It formally entered into force on 28 January 1988. Along with agreeing to participate in the EMEP data collection effort, Hungary committed itself through this participation to provide a modest level of financial support for the continued functioning of the system. More serious financing needs are involved in supporting the Hungarian-based part of the monitoring effort, and recently the Hungarian Meteorological Service, the government unit responsible for domestic data for EMEP, has experienced difficulty acquiring sufficient funding to gather the required data and insure its quality (Faragó and Horváth 1995).

Following the adoption of the EMEP system, the international acidification regime could address in more specific terms some of the major contributing pollutants that were the focus of concern for the acidification problem. The approach taken was to negotiate protocols serially, each addressed to a particular chemical (as opposed, for instance, to agreements based on an explicitly designated type of emitter/target group, like power plants or automobiles). A sulfur dioxide protocol, calling for a 30 percent cut in $SO_2$ emissions by 1993 (with 1980 emissions levels used as the baseline for comparison), was signed by Hungary on 9 July 1985 and ratified 11 September 1986. The protocol entered into force on 2 September 1987. Therefore, roughly a decade ago the LRTAP process resulted in the first of several specific and measurable commitments to controlling transboundary acidification on the part of Hungary and other countries.

The $SO_2$ agreement was followed by negotiations regarding another set of chemicals of concern, nitrogen oxides ($NO_x$). A protocol on nitrogen oxides was signed by Hungary on 3 May 1989 and ratified in 1991. In this case, unlike for sulfur dioxide, the full agreement distinguished among countries in terms of their targeted emissions goals and deadlines (see below). In the case of Hungary, the $NO_x$ protocol committed the nation to achieve by 1994 a level of emissions no higher than that prevailing in the base year of 1987.

Meanwhile, the LRTAP negotiations produced a third chemical-targeted agreement, this one on volatile organic compounds (VOCs). Hungary also signed this protocol, on November 19, 1991 (see Economic Commission for Europe, 1995, p.135). The agreement requires Hungary to stabilize VOC emissions at the 1988 level by the near 2000. The government has recently ratified this protocol as well (Ministry for Environment and Regional Policy 1995), after delays largely unrelated to the substantive issues at stake (although Hungary has not had reliable information on its nonmethane VOCs; see below) but connected primarily to the change in governments between the first and second coalitions of the post-socialist period.

A second, renegotiated $SO_2$ protocol has been completed at the international level and has been signed, although not yet formally ratified, by the Hungarian government as well. The signature date is 9 December 1994.[2] The commitment calls for reductions of 45 percent in $SO_2$ emissions by the year 2000, with 1980 as the base year; subsequent targets are also included: 50 percent (by year 2005 compared to 1980 levels) and then 60 percent (by 2010). Once again, somewhat different targets are established for different countries.

Representatives of Hungary are currently participating in the preparation of a revised $NO_x$ protocol, efforts for which began in Geneva in March 1994.

These last two protocol developments, it is useful to note, are based broadly on perspectives deriving from what is called a 'critical-loads' (or threshold levels) approach. Some analysts and national representatives favor moving away from single-chemical protocols and specific national overall targets, in favor of incorporating the concept of critical loads at the core of international controls. The critical loads approach would aim to target reductions particularly for those regions that are especially stressed by harmful emissions and would rely on computer modelling to identify and reduce the most harmful emissions. Furthermore, some experts advocate explicit consideration of interactions among the regulated chemicals (and also interactions with other potentially harmful chemicals that may be the object of control through other international agreements aside from those negotiated through LRTAP). Critical-loads principles can make major differences to the countries involved. And a critical loads approach, being complex and slow to develop, may have the disadvantage of providing political cover for those seeking to delay imposition of further restrictions. (The current Hungarian position on the issue is to favor technology-based standards during the transitional period to the eventual establishment of critical loads-based requirements.)

In addition to its commitments through LRTAP, Hungary is party to additional agreements and international ties. Of particular importance is the fact that the country is now an associate member in the European Union and is formally committed to achieve membership at the earliest opportunity. As a consequence, compliance with EU regulations is necessary and is receiving a high priority at present. This emphasis requires high-level national review of existing policy on the environment as well as other subjects and therefore is driving policy deliberations that may bear on LRTAP-related issues, as explained in a later chapter.

## Hungarian Governments' Approach to LRTAP

The process of Hungarian involvement in LRTAP has been one that reaches back to the early stages of the formation of this international acidification regime. In its own involvement in LRTAP negotiations, and the ways in which those negotiations affect actions within Hungary, the government has taken an approach that has remained quite stable over time - indeed, remarkably so, when one considers the range of changes that have taken place within Hungary since negotiations began on the first sulfur dioxide protocol in the mid-1980s.

The initial involvement of Hungary in LRTAP, and the kinds of perspectives the country brought to international negotiations on acidification, reflected both tensions between political blocs and alignments and competition in international economic trade.

First, a confluence of political interests encouraged East-West efforts at environmental negotiation by the mid-1980s. The United States expected to be able to place economic pressure on the Soviet Union by encouraging increased international attention to environmental protection, while the USSR's desire to initiate openings to the West and demonstrate international cooperation and leadership made the environment an attractive subject - in part to deflect outside pressure on human rights issues.

Meanwhile, international competition had developed for the marketing of investments in environmental technology. Japan was one nation in the forefront of such sales efforts, with some of the countries of the European Community also involved. Germany in particular had an interest in developing a share of such investments. Both Japan and Western Europe thus began to focus on the COMECON countries, in Eastern Europe, as possible centers for environmental investment. As (West) Germany became increasingly involved in acidification concerns and also environmental investments, its government initiated bilateral environmental assistance to Hungary, while also pressing for strict emissions limits for air quality purposes. Japanese inroads were thus limited. The overall development of the environmental management system was strongly influenced by the Germans. This point held even prior to the introduction of German technology on a large scale, but was further strengthened by investment in the 1980s.

In the end, Hungary's participation in negotiations on LRTAP was guaranteed by Soviet interest, as a relatively small country caught between superpower politics. And on some issues of significance within the acidification regime, Hungary has tended to side with the German point of view on how to address the needs for air quality (particularly a focus on best available technologies rather than reliance on the critical loads concept as a centerpiece).

At the same time, from the Hungarian perspective, dealing with transboundary acidification was a sensible course of action. The nation possessed significant scientific expertise and had been involved in international scientific exchanges for some time. Issues of environmental quality had become matters of increasing concern, given the conditions within Hungary (although acidification was not deemed the most important issue). In particular, public health concerns had arisen in some localities. And some analysts attribute Hungary's acceptance of the first sulfur dioxide protocol to a desire to improve domestic air quality, thereby addressing public health threats (Levy 1993, p.92).

In this context, the Kadar regime, which was in power at the time, was supportive in principle of the acidification initiative. On other issues of international cooperation in the recent past, Kadar and the political elite had taken strong 'top-down' positions in support of Hungary's making good faith efforts to abide by international commitments regarding data gathering, monitoring, and collaboration with other countries in joint enterprises.

After Hungary had joined the international acidification regime through LRTAP, for instance, but before any commitments had been made with regard to chemical-specific protocols, the state-socialist government - particularly the Ministry of Foreign Affairs - declared unambiguously that the nation should undertake obligations through international agreements only under conditions that made it highly likely that Hungary would be able to meet its requirements completely. This position subsequently influenced the approach and the process adopted for internal preparation for undertaking LRTAP protocol negotiations, as well as reconciling international and national decision making, and - to some extent at least - implementation within Hungary.

Hungarian negotiators, generally speaking, have gone into international negotiations for each protocol with the position that the nation wants to fulfill its international obligations; but if it is to do so, any agreement reached must take into account 'conditions within Hungary'. Of particular importance, of course, are domestic economic conditions.

By the time LRTAP had been initiated, and in part because of Hungarian interest in broadly international matters and the nation's strategic location adjoining the European Community, domestic specialists - unlike colleagues in some neighboring countries - had already developed connections in the West, including nations like the United Kingdom and the Netherlands. And Hungarian language facility in English, more well-developed than that in other COMECON countries, encouraged involvement in the international negotiations and networking in the newly-emerging acidification regime. The Hungarian LRTAP delegation was formed prior to those in other nations of Central and Eastern Europe, and Hungarian specialists played roles in the development of some important components of the new international agreement.

An example is afforded by the development of RAINS, the computer model devised at the International Institute of Applied Systems Analysis, in Laxenburg, Austria. This model, developed with some help from Hungary, was highly controversial at the time of its first introduction - and was used as a political tool to pin significant responsibility on the UK for transboundary acidification problems in Western Europe. The model is now much more widely accepted, even if not well understood outside technical circles, and continues to form part of the informational and technical infrastructure of the international regime.

**The Interministerial Steering Group**

To develop the specifics regarding the Hungarian position in each of the several LRTAP emissions protocols (sulfur dioxide agreements, the nitrogen oxide protocol and the renegotiations currently underway, and the agreement on volatile organic compounds, all to be discussed more carefully in Chapter 10), the government - through the ministry for environment - established an interministerial steering committee. This group includes representation from each portion of the government that might have an interest in the implications of an international agreement on acidification. Among the units included, aside from the environmental ministry and its background Institute for Environmental Management, are the Ministry of Finance; the National Institute for Public Health; the Ministry of Transport, Water Management and Telecommunication; the Ministry of Agriculture; and the Ministry of Industry and Trade. Certain individual participants in the interministerial group have been involved for some time - indeed, from the beginning of the LRTAP process. In many respects, the conduct of this group and its approach to the international negotiations and commitments have remained relatively stable over time, despite the major political, economic, and institutional changes occurring around it. In this respect, then, the Hungarian response to LRTAP mirrors its behavior in dealing with other international environmental agreements during the years of transition. (See Comisso and Hardi 1997 for evidence and conclusions regarding several other agreements that are quite similar to the findings for LRTAP presented in this study.)

For negotiations on some of the earlier protocols the former Ministry of International Economic Relations was also involved, but only in an observer status. (The change of governments in 1994 resulted in a reorganization eliminating this unit.) The Ministry of Foreign Affairs is, practically speaking, represented indirectly through the Ministry for Environment and Regional Policy, since its Department for International Relations, which collaborates closely with Foreign Affairs, is consulted during deliberations. The interministerial negotiations take place primarily among, and at the level of, technical experts. Therefore, it has been felt in the government that there is no need for a high-level foreign affairs representative. Of course, at the appropriate time later, for each internationally negotiated understanding, Foreign Affairs is responsible for signing the formal international agreement.

Some observers of environmental policy making in Hungary suggest that the relative noninvolvement of Foreign Affairs in such decision making reflects, as well, the lack of interest in environmental issues in other ministries and among those specializing in foreign relations and international affairs. In other words, the established pattern on LRTAP within Hungary is a manifestation of the secondary status of acidification and - in general - environmental policy in the national context.

Other observers offer a different interpretation, one based in bureaucratic politics. They point out that several years ago the Ministry of Foreign Affairs exercised considerable control over the development of international environmental agreements, largely because of the influence and connections held by one department head within that Ministry. Upon that person's appointment three or four years ago to head the country's permanent delegation to the United Nations, the influence of that unit (and Foreign Affairs) waned. The field became more open for others, and control was mostly absorbed at the Institute for Environmental Management.

What is clear is that the interministerial body has served as the forum through which the government's position has been concerted prior to and during negotiations for LRTAP protocols, as well as a significant center of discussion and analysis in the period after an international agreement has been signed, as indicated later. In their deliberations the representatives advocate the perspective of their own portfolios, while the agenda of the intragovernmental negotiations and the specifics of proposals considered and adopted are developed through the negotiating team, centered at the Institute for Environmental Management.

## The Institute for Environmental Management

This organization, formerly known as the Institute for Environmental Protection, was established in 1981 by the President of the National Authority for Environmental Protection and Nature Conservation. It was built from an earlier unit, the Institute for Air Pollution Abatement. During the period of state socialism, the Institute for Environmental Management was the so-called 'background institution' for the ministry for environment. Such units, common in state-socialist systems for various ministries, handled technical issues, regulatory and legislative drafting, research, and analysis for the ministries in question.

The political changes of 1989 have altered many aspects of social, political, and economic life in Hungary. But the Institute for Environmental Management has continued to operate, and some of its staff have played crucial roles in the LRTAP process. The Institute is no longer an official public body. After some debate and indecision in the period since 1989, the government decided that the Institute would become self financing. The unit now continues to conduct considerable work on a contract basis for the environmental ministry. And of course informal contacts and old friendships ensure that the Institute remains heavily involved in many environmental policy and technical issues. But a formal measure of independence has been established.

Indeed, the independence has meant difficult times for the Institute under current conditions in Hungary. Institute management has had to scramble for funding from multiple sources, and the budget has gone into the red. Cutbacks experienced by the Ministry have been felt in turn at the Institute, while the additional business generated from other sources - firms, consulting and engineering offices, and research centers - has not taken up the slack. As a result, some of the most able staff have departed to work for new firms in the environmental field or for other institutes and educational institutions.

However, until very recently the core of the LRTAP expertise has remained - particularly in the person of the Head of the Institute's Department of Environmental Consulting and Engineering, a long-time administrator in this organization until his move to the environmental ministry in early 1996. This Department has the largest of the three comprising the Institute, with approximately half of the total Institute staff of 90 during 1995. The individual occupying the role of Department Head until very recently has been the senior member of Hungary's LRTAP negotiating team. He participated on the team for every one of the emissions protocols, and he remains in place for the current nitrogen oxide renegotiations process. Throughout the entire period, he has occupied this official role - indeed, has served a term as chair of the Working Group on Technologies for the LRTAP regime - while also serving on the Institute staff. And it has been at and through the Institute that much of the drafting of national positions and the concertation of the interministerial discussions has been developed.

Of course political decision makers at the highest levels of government are hardly incidental players, either before or after the political changes several years ago. The nation's Council of Ministers in the current government plays the ultimate role in considering any prospective international commitment. However, the knowledge and continuity provided through the Institute, and the process of discussion and analysis centered in the interministerial steering group, have clearly influenced Hungarian decision making on acidification.

### The Steering Group-Centered Process

At every major step in international LRTAP negotiations, for every protocol, the interministerial steering group has met to discuss what the position of the Hungarian delegation should be and then to provide explicit guidance. The group also reviews progress on an annual basis and considers the need for new initiatives, as these may arise. The ministerial representatives provide the kinds of perspectives and analyses that might be expected, given the portfolios involved, and the Institute representatives remain influential through their drafting of background material, conclusions, and suggestions. Prior to the sessions of the steering committee, Institute representatives have typically tried to coordinate informally with the key actors from important ministries like Transport and Industry and Finance. (And these players, in turn, coordinate informally with others, including even representatives of industry. See below.) Questions of financing and also possible conflicts with programs of other ministries typically arise. The National Institute for Public Health is usually supportive of the 'environmental' position during these phases, an understandable point given the mutually supportive overlap in jurisdictions here.

When some level of agreement seems to have been reached, the Institute representatives typically draft materials for consideration at meetings of the committee.

During the ten years during which emissions protocols have been negotiated for LRTAP, there have been some adaptations - despite the remarkable stability of this portion of the institutional setting within Hungary. In particular, there is evidence that the evolution of the international agreement into a small but significant regime has assisted in the development of the Hungarian interministerial group into a more effective center of negotiations and coordination within the country. (The relationship may be reciprocal, as well.) Stability of tenure, the presence of needed technical expertise, and the relatively low-profile position of the group and the Institute within Hungary have probably assisted in this maturation as well. There may be costs, in terms of fresh perspectives and involvement of a broader public in LRTAP discussions. But certain advantages accrue as well.

The interministerial group has included increasingly experienced negotiators, specialists who have learned about the acidification regime and its larger international political and economic context, and ministerial representatives who have acquired over time knowledge of each others' perspectives and the needs and constraints faced by counterpart institutions of governance within Hungary. The group has come to expect that the international agreements will be treated seriously by other governments as well as their own, and they have developed working knowledge of the acidification regime and its strengths and limitations. Coordination within government and between government and the international level have improved (although coordination *by* government during implementation has faced serious gaps and challenges, as explained below). The improvements seem due, furthermore, both to adjustments in institutional expectations and also increased understandings at a more personal level. A number of individuals involved have served for some years and have learned to work well with each other despite policy differences. In these senses it would seem that the first protocols triggered Hungarian governmental processes at the prenegotiation stage that had influence during the same stages of later protocol negotiations. On the whole, participants in the process within Hungary characterize the efforts of the group as more effective and sophisticated, as the international regime and the domestic counterpart institution have evolved.

At the earliest stages (the first sulfur dioxide protocol), the steering group functioned somewhat clumsily, primarily because of the novelty of the activity. Different ministries sent officials from different levels in the hierarchy, and coordination and decision difficulties were experienced. Gradually, appointees to the group from the ministries became more nearly equal in terms of hierarchical level, thus facilitating within-group mutual consultation and decision making. The individuals became more clearly substantive experts who could be expected to actually conduct the work, rather than political figures from the apex of the bureaucracy.

Those close to the scene of the intragovernmental work at the prenegotiation stage thus characterize developments over time as a process of 'learning': about individuals involved, ministerial programs and needs, the significance of international agreements on the environment, and 'working out a good compromise', as one involved characterizes it.

Similarly, the non-environmental ministries involved in the interministerial bargaining have learned, according to observers and participants, that the protocols matter, that they need to receive attention, and that it is not in their interests to resist good faith participation. The issue no longer arises from any quarter.

Possibly the best measure of the stability involved in this phase of the Hungarian LRTAP experience is the following: other than the relatively modest shifts just reviewed, observers and participants note no significant differences in processes or outputs regarding the government's position across the time period in question. In other words, the 'system' at this level worked the same during the state socialist period as it has more recently after the political and economic changes.

Those involved point with pride to the nonpartisan nature of their efforts, indicating that the issues involved are considered important but they are treated seriously by partisans from all directions. Note, for instance, the signature in late 1994 by the newly-elected coalition government to the second sulfur dioxide protocol negotiated largely during the tenure of the predecessor governing coalition.

The overall picture, therefore, is one of 'learning amid a largely stable setting and process for negotiations'. In this regard, it is useful to mention one additional kind of evidence of learning on the part of governmental participants as they have gained experience with the process of prenegotiation prior to participation in international protocol negotiations. At the beginning of the LRTAP emissions protocol process, when preparations were underway for the first sulfur dioxide discussions at the international level, the amount and quality of communications were relatively low between relevant ministries and the various (at that stage, fully state-owned) industries important for emissions. Despite the fact that these were all formally part of the same state apparatus, the extent of coordination toward environmental objectives was low.

The prenegotiations planning, therefore, for the first couple of protocols was somewhat difficult and less than precise. Those involved in this experience, however, recognized the value of information gathering and communication with sectors involving important emissions sources; and through the relevant ministries the core group began to develop coordinative ties with certain industrial parties. As it happened, some of this earlier effort initiated under the one-party regime could be used to the government's advantage even after the political and economic transformations of 1989-1990.

In fact, it is useful to note, the period under state socialism had offered two sides of the coordination coin. (Note the differences in coordination dilemmas experienced in a state-centered system like the one prevailing in Hungary prior to 1989, on the one hand, and multiparty arrangements with significant private control over investment and production, on the other. See Hanf and Underdal 1996 for an ostensibly general treatment that focuses implicitly on the latter.) On the one side, the state-centered system formally allowed an 'easy' avenue for coordination via adoption of a general directive, subsequent to which all organs of the apparatus would be obliged to obey. In fact, even during the period of initial experience with LRTAP emissions protocols, the government officially required that any such agreement be 'harmonized' with those who would actually be subjected to it. However, in practice, if the industrial sector - or even a particular firm - was sufficiently influential (via personal contacts and lobbying), the procedures and even applicable sanctions were subject to negotiation or exemption. Thus, for instance, the power and cement industries were exempt from the relevant air quality standards under the emissions controls placed into force by the government at that time.

However, beginning in the late 1980s, as government gradually let go of direct control over some of these key industries, it became increasingly clear that efforts would have to be made to link the official negotiations with the possibilities and planning requisites of the newly-emerging private sector. The ministries involved in the prenegotiations process recognized that they needed to coordinate with some of these key industries if there was to be any hope of executing the protocols in practice.

As a consequence, the Ministry for Environment and Regional Policy in particular began to undertake informal negotiations with key elements of major industrial sectors to elicit agreements *ahead of time*, that is, prior to the protocol negotiations. In this fashion, the negotiators and the interministerial screening group acquired a much better sense of what would be possible, and probable, with regard to emissions projections - at least regarding industry. (Obviously, for some pollutants, vehicle emissions and emissions from households and small businesses are also important parts of the equation. Note the $NO_x$ data presented in Chapter

4. And increasingly, as smaller industrial firms also become active in Hungary in the wake of political changes, the relatively limited informal communications channels have proved less satisfactory. This last issue is treated below, as a part of the institutional requisites for implementation during and after the period of political transition.)

According to those involved in several protocol processes, there has been 'no problem' with acceptance and compliance from industry with moderate requests for cooperation, so long as the latter had been involved at earlier stages, devoted time and energy to the prenegotiations (thus involving sunk costs in the course of action), and had come to accept the rationale underlying the overall approach and its meaning for them. 'If the preparation is wrong', however, disputes tend to arise later. As suggested earlier, the negotiators' view is that the quality of the decision making process at these early stages improved after the first $SO_2$ effort.

One should not make too much of these pre-protocol efforts to coordinate. As explained later in the protocol-by-protocol review, massive behavioral changes and emissions cutbacks cannot be traced to protocol-stimulated efforts. The evidence on this point is clear for the first enactments regarding sulfur dioxide and nitrogen oxides, and it is as yet too soon to offer definitive evidence for the more recently-approved agreements. However, some concertation has been attempted, and it is evident that some changes have been induced in the industrial sector.

At a minimum, the efforts have resulted in information sharing and some strategic telegraphing of likely constraints for the future. It is reasonable to conclude that these efforts have lowered the level of conflict on air quality issues in the difficult transitional environment. And there is some evidence that anticipated international restrictions, communicated to some state industries, have become incorporated into the planning processes in these sectors. Therefore, although Hungary has not yet developed explicit, governmentally endorsed emissions targets for the country - outside of the formal commitments to the LRTAP protocols - the implications of these commitments became significant inputs into decision making in the management of certain key sectors.

An example of the coordination achieved with certain industrial sectors is provided by the Hungarian power industry. The government, including the Ministry of Environment, has had longstanding connections with this sector of the economy. Indeed, the government, and in particular the Institute for Environmental Management (now no longer a formal adjunct of government) has performed contract work for the industry regarding technical possibilities for reducing emissions, financial aspects of such technological changes, and so forth.

The Hungarian Electricity Board, responsible for developing plans for this portion of the industry, began to develop a long-term energy plan. This group took the initiative to approach the ministry for environment to determine what their requirements would be for periods of five to ten years into the future. As a consequence of this coordination effort, the electricity industry began to adapt its production to emission limits that have yet to go into effect. And, according to insiders familiar with this kind of ministerial-industrial networking, the prospects for coordination and compliance by large industry with governmental policy have been enhanced by the economic changes initiated several years ago.

The combination of the force of the international agreements, once negotiated, and the bargaining among ministries and with major industrial sectors during prenegotiations provide some leverage to the government, and especially to the ministry for environment, to extract the needed actions from the range of interdependent actors. Of course, the Hungarian political and economic systems are in flux, and in any case there is nothing approaching a corporatist-like understanding in the shifting and evolving private sphere or public industrial realm, so the changing institutional understandings and relatively limited range of control by the environmental ministry tempers this coordinative reach in practice. But there are institutional connections which matter in practical terms in connecting the several levels of action.

To explore how these processes have functioned as the LRTAP protocols have been developed and adopted, it is useful to consider in depth how Hungarian actors have sought to influence the international negotiations and, in particular, how the products of the acidification regime have affected policy and the policy process within Hungary. Chapter 10 addresses these issues.

## Notes

1 In addition to these structures, other institutions providing background materials to assist in government decision making on LRTAP protocols include EGI (the Energy Institute), MVMT (the

state electricity company), ERTI (the Forestry Institute), and TAKI (the Soil Institute).

2 Hungary signed in December, rather than months earlier, because the international agreement was reached at a time of transition between the outgoing and incoming elected national governments within the country.

# 10 Protocols into policies?

The processes of negotiations, policy making, and planning for implementation within Hungary have involved intricate coordination efforts. These have been concentrated primarily around a fairly small set of actors, during the state-socialist period and also in the years of political and economic transition. The processes have also yielded an array of impacts, both obvious and subtle. Yet, as the coverage in this chapter and the next one makes clear, growing challenges in policy making and - especially - implementation suggest an imposing set of difficulties that will require significant adaptations to meet the nation's international commitments in the future.

This chapter first presents information on how negotiated LRTAP protocols have been converted into domestic policy and plans for action within Hungary. Then detailed analysis is introduced of the principal protocols and what has happened within Hungary as a result of their adoption. While some data regarding implementation are presented in this chapter, a more thorough review of this topic is provided in Chapter 11, where particular attention is given to the trials of execution during the years of political and economic transition.

## LRTAP Protocols and Domestic Policy

As protocols are being negotiated, the Hungarian participants must decide how, if at all, to adjust their own bargaining positions; and whether to recommend acceptance of the final result to their government. Of course, consultations with the affected sectors are ongoing, as represented in the interministerial group. And ultimately the nation's Council of Ministers must make the formal decision in each case. Some information on these elements of the Hungarian process are contained below, in coverage of specific protocols.

These aspects of the process, nonetheless, provide only part of the story. Once protocols have been negotiated and assent given by the Hungarian representatives, the team involved then undertakes a process of negotiating within the government to encourage ratification of the agreement within the country. During this phase of the process, as well, the Institute for Environmental Management staff have been centrally involved, working especially closely with the environmental ministry to translate the stipulations of the international protocol into an explicit national commitment - albeit one that has not been stated in terms of national quantitative plans for reductions, specific programs that have been initiated solely in response to the international agreements, or other easily measurable and monitorable features, as explained shortly. Part of this process means elaborating the protocol's meaning in the context of Hungarian institutions; policies; and social, political, and economic circumstances.

All these basic elements of the domestic picture have remained as relative constants since the first sulfur dioxide protocol in the mid-1980s. Therefore, these aspects of the process can be discussed broadly in the next few pages, with some attention as well to the important question of whether the efforts make any practical difference in changing behavior. This coverage is then followed by information on the individual protocols.

In general, the process within Hungary of attempting to convert the requirements of LRTAP protocols into domestic decisions involves integrating a particular protocol's targets for Hungary with the national strategy for dealing with emissions and air quality. This strategy and its underlying policies heavily emphasize human health concerns. The Hungarian perspective strives for integrated, coherent policy in domestic terms, not policy driven primarily or inordinately by international agreements.

In broadest outline, the national strategy has contained three central tenets:

- Air quality in heavily polluted areas in several parts of the country should be improved.
- In those portions of the country where air quality is relatively high, this state of affairs should be maintained.

- International agreements affecting air quality should be a forum of participation by the Hungarian government, and the results of such agreements should be implemented. Hungary, recognizing that its air quality is heavily influenced by actions in other countries and that its own policies and practices have significant impacts on environmental conditions abroad, seeks to act on the basis of the premise that national air quality can only be enhanced to the extent that the nation participates in international agreements in Europe.

From these three injunctions can be deduced corresponding programmatic efforts: to define boundaries for and focus targeted efforts on heavily polluted areas (with efforts in turn geared to the most significant emissions sources in a given area: industrial, vehicular, and/or residential); to protect those areas which now enjoy relatively high air quality by such measures as preventive actions, for instance the thoughtful use of environmental impact assessments; and to participate in international forums so as to link international commitments with national needs and standards.

The results of this sort of approach, when combined in recent years with the economic changes underway, the more separate roles between government and industry leading to less cozy but more productive bargaining, and the increasing internationalization of the Hungarian economy, can be seen in the process during the prenegotiations for the second sulfur dioxide protocol, as explained later in this chapter. The general picture is one of behind-the-scenes discussion and information sharing, rather than an explicit public adoption of programs aimed at dramatically changed behavior.

This characterization might seem to suggest predominantly negative evidence regarding the impact of protocols themselves on the behavior of parties within Hungary. One perspective might be that, for instance, industrial changes that emerge in the wake of consultation (as in the coverage of the Hungarian Electric Company and Hungarian Oil Company, below) would have occurred in any event; therefore the international negotiations have been and are likely to continue to be irrelevant for actual results as observed in Hungary. When the position of the government and the narrowly-construed interests of large firms are in concert, the argument could be, the commitment of a nation like Hungary to international agreements is irrelevant or, at most, an epiphenomenon of the economic forces at work - at least in recent years during the time of increasing market presence and pressures.

However, evidence in the Hungarian case suggests that this interpretation understates the influence of the international processes on intranational action (as well as the influence of Hungarian experience on its positions in international arenas). In particular, several of those involved in the concertation process point out the felt need on the part of both government and industry for collaborative discussions and some degree of both bargaining and coordination. The process is perhaps best described as something apart from a pure coordination game (in game-theoretic terms) and also something quite different from a zero- or negative-sum game.

Governmental representatives, explicitly or implicitly, seek to protect Hungarian industry and 'national needs on economic matters' at the international level; but seek, in return, some acceleration of pollution abatement efforts and some coordination of investment plans with international protocol needs.

As mentioned earlier, these different arenas and the considerable but not perfect overlapping of interests are understood well among governmental officials involved and also representatives of the largest firms. The matter is less clear, less coordinated, and operates less well for the smaller firms (see below).

Obviously, these three broad targets are interdependent and cannot be achieved without explicit consideration of how they can best be approached in an integrated fashion.

Part of that coordination process occurs in the prenegotiations process explained above, as national priorities and broad strategies are articulated both 'downward' toward firms and major sources of emissions and also 'upward' toward feasible and acceptable international negotiating positions. Part of it sometimes does not come at all, as the internationally- and nationally-made decisions are attenuated through the complex implementation pattern that constitutes Hungary's air quality programs (see below). But partly, the link between transboundary air quality protocols and the results in the world of practice are made, or at least attempted, via the translation of the internationally-negotiated targets into explicit national decisions enacted after the protocol's completion; and also via the conversion of such national decisions into streams of policy-derived action in the field.

More specifically, the Hungarian process following enactment of LRTAP protocols has been roughly as follows. Once a protocol has been accepted, the government - again with the environmental ministry and core of acidification experts at the Institute for Environmental Management - develops a list of relatively specific tasks to be accomplished and incorporates these into a document called a governmental decree. (For a recent example on air pollution, see Governmental Decree 1079/1993 (XII.23)Korm., covering actions needed for the period 1994-1998.)

These activities are not detailed enough to constitute a real implementation plan. Rather, the decree specifies streams of action such as signing the international protocol (a task assigned to the Ministry of Foreign Affairs), developing legislation, writing action programs, tasking certain units with designing the nuts and bolts of implementation processes. In each case, specific governmental ministries and units are identified as the responsible parties, deadlines are specified, and some allocation of tasks is effected. The decrees have the status of draft 'strong instructions' to the relevant parties, and there are always at least some items included that refer to initiating the implementation of the task.

For instance, the decree may assign the job of developing a detailed implementation plan to a particular unit of government, generally the ministry for environment. And this task, in turn, generally has been assigned (nowadays, by contract) to the Institute for Environmental Management. The implementation plan itself may include a long list of more specific tasks; the inventory may approach one hundred items. Eventually, a version of this plan is accepted by the Ministry, which then seeks approval from the government for proceeding with action.

If one were to seek to examine these implementation plans to determine just what tasks and behaviors flow, and have flowed in the past, from international protocols, one would be unable to reach clear conclusions. This point follows from the process and form by which the implementation plans are designed.

The staff work done at the Institute and in the environmental ministry integrates the protocols, or more precisely the interpretation by the key staff specialists of their meaning in Hungarian terms, into the general tripartite national strategy mentioned above. The protocols and the specific actions they trigger disappear from overt view beyond the level of the governmental decree. They are not mentioned in the implementation plan, and the specific behaviors included as draft mandates in the plan are not articulated in any measurable way back to the quantitative reductions or freezes that appear in the international protocols. The 'connections' appear in the calculations of the experts, and the cognitive maps of the central decision makers, as they interpret how they expect to be able to effect specific influences in aggregate emissions across sectors from the proposed set of actions.

In the present investigation, efforts have been made to undertake the difficult task of assessing indirect evidence of the impact of LRTAP agreements by gathering information *ex post facto* within Hungary. Two principal data gathering efforts along these lines have been attempted: (1) content analysis of the decree itself and certain elements of the implementation plans, enacted in the post-protocol periods, for evidence of implicit LRTAP influence; and (2) structured interviews with knowledgeable elites and participants to gather their *ex post facto* assessment of acidification regime effectiveness for particular protocols.

With regard to the former line of inquiry, it has been very difficult to identify evidence of influence. There are two reasons. To the extent that information is available about them, the decrees and plans leave much unsaid, and major new initiatives cannot be traced unambiguously to protocol requirements. The 'causal theory' (see Mazmanian and Sabatier 1990) underlying the government's approach is by no means fully specified in terms that allow for systematic review and assessment. Also, some of the specifics of the implementation plans have not been made available for independent examination for this project.

With regard to interview evidence on *ex post facto* acidification regime effectiveness, some specifics are reported below under discussion of particular protocols. Several knowledgeable elites and participants have been asked for their own quantitative estimates of how much, if at all, the emissions targeted in LRTAP protocols would have been reduced during the designated period as incorporated in the international agreements, had the agreements not existed at all. This assessment of the counterfactual obviously cannot be independently verified, but the responses provide a rough measure of the impact of this international acidification regime on particular air quality problems. Overall, despite the confounding dynamics alluded to early in this study and discussed in more detail below, those interviewed for this issue do estimate some independent impact contributed by the protocols. Needless to say, none of those queried believes that the full extent of Hungarian emissions reductions can be attributed to the international agreements.

Despite the inevitable ambiguities, therefore, it would be an error to conclude that the LRTAP protocols do not have any practical meaning. There is some perceptual evidence for overall impacts. In addition, information highlighted by tracing the processes involved in approving the protocols in Hungary and implementing air quality programs during the period of LRTAP's operations suggests that, while substantial implementation gaps and additional influences complicate the picture, the existence of the international regime and its agreements have exerted some influence on behavior and emissions in Hungary.

A clearer understanding of this modest but likely influence can be developed from some brief treatment of individual protocols. Particular emphasis is given to the earliest protocols, since these have probably been more important in influencing the processes and outputs at the domestic level regarding acidification matters, and they have been operational for a sufficiently long time to permit some assessment of impacts, both short- and long-term.

## The First Sulfur Dioxide Protocol

The first $SO_2$ agreement through LRTAP was the initial attempt to control acidification in specific terms through the establishment of numerical targets for emissions reductions. Hungary, along with other signatory nations, had to consider what kinds of cutbacks would be feasible under the circumstances anticipated at the time of the mid-1980s. The $SO_2$ protocol process was also the first experience developing the interministerial steering group and using the Institute for Environmental Management as a focal point for organizing national efforts.

Discussion at the international level focussed on the target of a 30 percent reduction on the part of all nations participating. Hungary had relatively little innate concern with transboundary issues at the time but was motivated to consider an $SO_2$ commitment seriously because of desire within the diplomatic community for cooperation with the West, and also domestic concerns about public health within the country (see Levy, 1993, p.92).

Within Hungary, there was no policy at the time targeting any level of emissions; and no real experience planning for the achievement of targeted reductions like this. Two key features of the Hungarian situation were well known at the time, however. One was that several processes either already agreed upon or underway would have the effect of reducing emissions of sulfur dioxide. The other, a subject of discussion among the negotiators and in the interministerial group, was that the selection of 1980 as a baseline would be advantageous from a domestic perspective. It was clear to analysts that that year represented an emissions peak from which reductions had begun and would be almost certain to continue.

For $SO_2$, energy consumption would have to be a key in any prospective emissions reductions. In the decade of the 1950s Hungary had taken preliminary steps to increase energy efficiency, but the trend reversed during the next two decades with large supplies of coal and oil available. A shift was undertaken from coal to oil to increase independence from foreign suppliers, but the oil crises of the late 1970s reversed this trend. By the early 1980s, the nation had embarked on energy rationalization programs and begun to produce savings again. 1980 thus provided a maximum of leeway for any given proportionate reduction enacted. The international agreement on 1980 as the baseline selected was therefore highly advantageous from the Hungarian viewpoint and made the decision regarding support for the protocol significantly easier.

Still, the decision to sign was not undertaken lightly, as implied earlier in this study, and involved heated political discussions not only in the interministerial group but at the highest political levels of government. Once the government decided to support the protocol, the issue took on significance within governmental circles influenced by the top-down commitment to meeting internationally-negotiated agreements. In the case of this protocol, the goal was thus to reduce $SO_2$ emissions by 30 percent, from the 1633 kilotons recorded for 1980 to no more than 1143 kilotons annually by 1993.

The ministries represented on the interministerial steering group considered and debated among themselves the implications and possibilities for reductions. In this first protocol process, the practice began of ministerial analyses, supplemented by work at the Institute, to project trends and identify options in response to the negotiated efforts.

It became clear quickly by examining available data and gathering additional information from the energy sector during the 1985-1986 period that certain decisions already undertaken would help in reducing the emissions levels, and other commitments or events stimulated by causes or motives aside from environmental protection also were to have favorable impacts on the emissions totals. In particular, shifts in the fuel mix, structural changes in industrial sectors, and ultimately the serious economic recession developing near the end of the decade all helped the emissions picture. The projected opening of a nuclear power plant and plans for increased use of natural gas for home heating (from domestic sources, and also the USSR via a new pipeline constructed in the early 1980s) were expected to be important. The recession could not have been anticipated, of course, but the other changes were already underway (see Ministry for Environment and Regional Policy, 1991a, pp.20-21).

Despite these projections, considerable uncertainty remained. The government had very incomplete information regarding transportation and projections for the future, possible changes in home heating patterns, and other issues. And no rigorous cost-benefit analyses were conducted at the time of the ratification decision, primarily because of lack of experience on matters pertaining to emissions cutbacks.

While projections at the time of the protocol's ratification in Hungary suggested reasons why the goal could be feasible, the best estimates at the time projected a shortfall if no additional action were to be taken. The trends underway suggested the likelihood of $SO_2$ cutbacks at the level of 25 to 26 percent by 1993. Options were considered by the interministerial group and technical specialists, particularly the industrial and environmental ministries and the Institute. Plans were developed to introduce flue gas desulfurization in some of the thermal power plants to reduce emissions from coal with high sulfur content. (This analysis thus differs from that in Levy 1993, which implies no Hungarian emissions impact from the protocol.)

Soon, however, the economic changes dramatically altered many trends, including those regarding emissions of pollutants like $SO_2$. The overall level of economic activity declined for several years, so that by the time of the 1993 protocol target date, Hungary's nationwide sulfur dioxide level had plummeted to 757 kilotons, thus reflecting nearly a 54 percent cut.

All parties involved realize that this huge reduction is due largely to the recession. Indeed, government decision makers continue to identify steps that can be taken to reduce $SO_2$ emissions in the future, particularly in light of both the second protocol on sulfur dioxide and the projected increases in economic activity likely in the next few years. For instance, fluidized bed combustion methods are planned for large-scale introduction in power generation facilities (Ministry for Environment and Regional Policy, 1991a, p.20). Assistance from the World Bank has been used to reduce sulfur dioxide emissions through technological changes at a major refinery. And the multiyear action plans begun in the wake of the first $SO_2$ protocol continue to be developed, with particular emphasis on improving air quality in heavily polluted regions. As indicated below, for instance, the third plan, covering the period 1994-1998, is designed to anticipate the need for additional reductions and encourage their incorporation into planning by important industrial sectors like energy.

However, it is fair to indicate that the $SO_2$ issue has not taken on a high priority since the onset of the economic difficulties. The accumulation of additional international obligations, including those associated with LRTAP but also other agreements (for instance, on climate change) and the great and increasing salience of harmonization with the European Union, plus the severe distress occasioned by the recession and the recent reductions in social welfare support, have pushed sulfur dioxide emissions far down in the struggle for attention and action by policy makers. Some environmental advocates and NGOs argue with considerable force that the transitional period for the economy is precisely the best time to make the investment and other shifts necessary to improve air quality for the long run. But attention has been directed primarily at other issues, for understandable reasons.

It is clear from the evidence related to the first sulfur dioxide protocol that the international agreement had impacts within Hungary. The protocol stimulated the congealing of certain institutional arrangements and the formation of a process for addressing acidification issues. Experience with the $SO_2$ requirements provided negotiators with an understanding of where the national advantage lay on acidification issues and began a process of determining where the most important domestic sources of transboundary pollution were located. (The importance of local effects for transboundary flows was documented as part of this analytical effort.) Interministerial communication and negotiation was initiated on an important issue. And linkages with key industrial sectors on environmental questions were clearly catalyzed in part through the LRTAP process.

Did the protocol influence emissions? The weight of the evidence suggests that the answer is: yes, but not nearly to the extent that the subsequent reductions might suggest to the superficial observer. A major portion of the negotiated cuts were already likely to occur on the basis of decisions made on other grounds. And the economic difficulties of the early 1990s 'contributed' another huge cut in $SO_2$. Estimates made in 1996 by insiders in the negotiations and policy-making processes in Hungary are that, in the absence of the LRTAP protocol, the 1980-1993 cutbacks would have been in the range of 25-35 percent. Given the achieved reduction of 54 percent, and even granting that these estimates might be mildly biased toward LRTAP's effectiveness, it is reasonable to conclude that the international agreement has had institutional, process, output, and probably outcome effects with regard to sulfur dioxide specifically, and in indirect fashions on the seriousness with which international agreements on the environment are treated.

## The First Nitrogen Oxides Protocol

The process sketched above for sulfur dioxide was used in general outline in the case of the first $NO_x$ protocol as well. The international agreement in this case was to stipulate that Hungary would be expected to meet by the end of 1994 an emissions 'freeze': annual $NO_x$ totals were not to exceed 280 kilotons per year, the 1987 baseline.

There were, however, some differences. First, upon the government's decision in 1985 to participate actively in the negotiations process for $NO_x$, a Hungarian representative was chosen to be directly involved at the international level in organizing the $NO_x$ effort, as chair of the LRTAP working group that developed the protocol proposal. Clearly, Hungarian participants were well-apprised of developments during the negotiations period. And accordingly, there was interest within policy circles in the government to press for Budapest as the locale for the signing of the protocol. (The arrangement of this event in Hungary was ultimately not possible, since upon reviewing its final features the government delayed in its decision about whether to sign. The main reason had to do with the financial implications of the agreement as regards steps needed in Hungary. See below. The protocol signing site was established as Sofia, since the vice-chair of the Working Group was Bulgarian.)

Second, whereas possibilities for $SO_2$ reductions were closely tied to power production within the country, $NO_x$ was related to transportation.

(Automobiles carried only 40 percent of the total transportation load at the time of protocol deliberations, but this fraction was to increase in later years.) Accordingly, the industry ministry played a less important role in ministerial deliberations and projections, while transport was more central in this case.

And third, since the apparatus for considering acidification issues was already established and had functioned for sulfur dioxide, the processes of deliberation and analysis were somewhat easier. One practical result was that those participating in the domestic assessment of $NO_x$ possibilities were able to conduct analyses during the period of international negotiations to determine the feasibility of *reductions* in emissions and concluded that trimming would be, from their perspective, prohibitively costly. The Hungarian position developed in opposition to cutbacks in nitrogen oxides, primarily because of the anticipation of growing production, plus projections of an increase in vehicles and the persistence of outmoded automobiles operating in the country (Levy, 1993, p.95). The domestic group instructed their negotiators to press for a freeze at 1988 levels.

Even with a cap rather than reduction as the goal, the Hungarian participants were uncertain how the required shifts in expected trends would be effected. Through the apparatus of the interministerial group, the issues were considered. Negotiations were particularly active and detailed among a subset of participants representing the environmental and transport ministries, along with the finance ministry.

In the case of this second protocol, the earlier experience and the increasing sophistication of the domestic analytic effort meant that clearer understandings of the implications of the international agreement were available. The transport ministry felt confident in being able to assure a freeze within that important sector, under the assumption of constant transportation levels. However, the growth in use of vehicles was rapid at the time, approximately ten percent per year. Meeting the protocol target under these circumstances seemed highly uncertain; and technological-regulatory fixes, like prescribing the use of emissions-reduction equipment on cars, was deemed infeasible - an expensive and unpopular option.

In particular, those closely involved in the projects determined that a relatively small increment of $NO_x$ - ten to 15 kilotons - posed a particular problem. To meet the protocol requirements by 1994, the group determined that this amount of additional cuts would have to be made, beyond what was projected through some of the shifts already underway (particularly those outlined above for the sulfur dioxide protocol). This relatively small additional reduction would cost several billion forints.

The government would not sign the protocol until concrete projections could be developed demonstrating the likelihood of success in meeting the negotiated LRTAP requirements for nitrogen oxides - thus the delay, mentioned above. To deal with the issue, those centrally involved (ministries represented in the interministerial group, a few environmental specialists, and the background institutes for environment and some of the other ministries) considered a broad array of options and eventually identified some combinations of initiatives and small shifts that seemed to be viable.

Meeting the goal was not expected to be costless. Importantly, however, international/acidification regime pressure moved the government at the political level to commit to the agreement.

Once again, of course, the major economic changes mooted the concern about compliance within short order. And the accelerated influence of market forces created other significant shifts in the transportation pattern.

Economic decline produced emission cuts beyond those projected at the time of deliberation over the $NO_x$ protocol. An increasing proportion of the emissions load for this pollutant derived from transit. And while the environmental ministry developed a proposal for consideration by the Council of Ministers in 1990, the press of the dramatic political and economic changes pushed nitrogen oxides out of the limelight. The proposal, focussed primarily on automobile emissions, was not even discussed that year (Ministry for Environment and Regional Policy, 1991a, p.21). Eventually additional transit measures, some essentially unrelated to $NO_x$, were enacted. The 'green card' system for annual automobile inspections was inaugurated in 1992. And a small tax on gasoline was initiated, primarily for revenue raising purposes (for the Central Environmental Fund).

In the meantime, vehicle use, which had been increasing, was cut somewhat - primarily because of the steep increase in gasoline costs upon shifts in energy pricing toward market rates. Vehicles made in the West and equipped with catalytic converters were introduced and have become a slowly increasing portion of the vehicular total. (Older used cars have also been imported.) The old two-stroke engines that have been commonly used in Hungary cause extensive $NO_x$ difficulties, and individuals facing economic hardship have generally been reluctant to give up their aging vehicles for expensive new ones, despite modest governmental incentives for retiring heavily-polluting cars. Therefore, the average age of vehicles driven in Hungary has increased. Thus two contrasting trends - gradually cleaner cars and increasing transit levels as economic conditions improve - can be projected for the future. Grappling with the issue in the transport sector, therefore, is by no means a *fait accompli*; and several small initiatives recently or currently being implemented are aimed at managing the nitrogen oxides problem. Some of these are mentioned below in connection with the second $NO_x$ protocol.

Again, then, the question can be asked: did the protocol in this case make any difference? Again, the answer would seem to be yes. The data presented here suggest that Hungary was involved early and earnestly in the international negotiations, the government treated the issue of committing to the protocol very seriously, pressure from the international level influenced the decision to sign, acidification experts and analysts in the domestic setting from several governmental units played important roles, and some program options were selected to alter the emissions trend line toward compliance - despite projected costs at a time of financial stringency. More sectors had become involved in responding to acidification issues from the international level, and the level of competence in analyzing and choosing options on these questions had improved from the earlier level.

Did these changed behaviors also influence outcomes? Again, the answer seems to be positive, although the overall quantity of $NO_x$ reductions is much greater than could ever be attributed to efforts induced by commitments to the acidification regime. The protocol called for a 1994 freeze at the 1987 level. Official 1994 figures are not available yet in published form. However, it is clear that the country has easily met the protocol standard, with room to spare. The emissions total for 1993 was 184 kilotons, a 34 percent reduction from the baseline. The large causal forces were the economic changes and the related shifts in the transport sector.

*Ex post facto* assessments by decision makers involved during this period estimate the reductions would still have occurred, but at the lesser levels of ten to 20 percent.

**The Protocol on Volatile Organic Compounds**

Hungary participated in the protocol negotiations on VOCs and signed the result in late 1991. Again, the familiar process was used in developing the domestic negotiating position, deciding about support, and considering options for implementation.

In this case, however, two important differences have affected the actions taken. First, the deliberations and decisions took place amidst the economic recession and under the authority of the post-socialist government. Therefore, while the institutional arrangements for acidification involving the interministerial group and the Institute were stable and consequential, the governing context and economic conditions so important for determining emissions levels and options had changed radically.

Second, unlike the situation for sulfur dioxide or nitrogen oxides, there were few reliable detailed data on the sources of VOC emissions in Hungary, prior to the period of negotiations.

These differences meant that the more experienced analysts and negotiators were aware that they were deliberating in a much more uncertain context, particularly with regard to future implications of economic growth. The government participants were seriously interested in achieving reductions but had little knowledge of what such an effort would mean, particularly under the economic circumstances. As a result, they argued for a freeze (see Levy, 1993, pp.99-100). The protocol as negotiated established 1988 as base year for the Hungarian commitment to a VOCs freeze targeted for 2000.

On the one hand, it was clear that the dire economic conditions virtually guaranteed compliance for the next few years. A substantial cushion had been created even in advance of any efforts to change behavior to reduce emissions. On the other hand, over the longer term and with the eventual stimulation of economic growth, the VOC cap could prove very difficult to meet. How severe a problem for the longer term was itself a question difficult to answer, given the absence of reliable data about emitters, options, and trends. Experts estimated the uncertainty in the available data as approximately 30 to 50 percent. And authorities had developed no experience dealing with some of the emitters that are likely to be important in any VOCs reduction efforts.

Indeed, some of those involved in the protocol deliberations felt that eventually it would be determined that this protocol would pose great compliance difficulties for Hungary and that signing the agreement would commit the country to an unreasonably large burden.

The history of participation, however, and the 'breathing room' provided by the recession provided sufficient motivation at political levels for ultimate agreement. Since the early 1990s, then, the technical experts and ministerial representatives have been seeking to deal with the implications.

Accordingly, the first concrete steps taken in response to the protocol have involved surveying the relevant industries to determine current practices and emissions. This effort was completed in 1993. Those involved in identifying options are now seeking to obtain support from the Central Environmental Fund for efforts to decrease VOC emissions. And bilateral support from the Dutch government has been directed at developing an emissions reduction plan. The aim is to implement an integrated plan with detailed reductions for each source type on a carefully timed schedule.

Scenarios have been developed to project VOC levels under a 'business as usual' scenario as well as on the basis of new measures that may be introduced. The uncertainty behind these estimates is approximately 15-20 percent. Implementation has begun but is in comparatively early stages.

Obviously no assessment can be made at this point regarding compliance with this protocol. It can be observed, however, that again the protocol requirements are being treated seriously (indeed, are a source of concern presently for the officials involved) and that streams of action have been initiated with the aim of matching commitments with achievements. The involvement of a bilateral aid program to assist with planning and implementation on VOC reductions suggests an increased likelihood of notable adjustments into the future. Yet the difficulties of projecting economic trends and industry practices regarding VOC implications are considerable. And the tough commitments, including some important spending decisions, have yet to be made.

In *ex post facto* assessments of regime effectiveness, some senior participants in acidification decision making estimate that in the absence of the LRTAP protocol on VOCs (including the absence of international agreements directly affecting countries besides Hungary), Hungary would not have reduced its emission levels for this class of chemicals between 1988 and 1994. They judge that a combination of increasing use along with difficult economic conditions would have kept the emissions relatively stable during this period. Once again, 1994 figures are not yet officially available. However, if one compares the 1988 and 1993 VOC emission levels, a decline of 30 percent can be noted, from 205 to 143 kilotons annually. Based on these figures, and even given the probability that participants in the domestic efforts to control acidification are likely to overstate the impacts of their efforts, it is reasonable to conclude that the LRTAP protocol initiative has had some effect on VOC emissions in Hungary.

## The Second Sulfur Dioxide Protocol

During 1993 and 1994, while in the midst of severe domestic economic difficulties and during the time of a national domestic political campaign (and transfer of power from one coalition to another), the Hungarian government participated in the international negotiations leading to a new sulfur dioxide protocol. For Hungary the resulting new emissions targets represented large additional commitments beyond the 30 percent involved in the first protocol. The dramatic drop in the $SO_2$ emissions levels during the 1990s, nevertheless, has meant that the country has already reached approximately the reductions necessitated for many years into the future (45 percent by the year 2000, with a 1980 baseline, 50 percent by 2005, and 60 percent by 2010; the reduction achieved by 1993, as reported above, was 54 percent). Retaining 1980 as the base year was important for Hungary; nations in Central and Eastern Europe argued successfully in negotiations that it would be unfair to expect them to be able to produce economic growth and trim $SO_2$ levels from those prevailing during the severe economic downturn. A result is that, even with moderate economic growth from now until the year 2000, no new measures are expected to be needed to meet protocol requirements. (This point ignores the possibility that negotiations on the second $NO_x$ protocol may add a combined critical load constraint to the individual-pollutant restrictions prevailing so far, thus possibly implicating $SO_2$ levels in an overall limitation.)

Indeed, the recession saw especially sharp drops in production in those industries (like metals and bauxite) that rely extensively on energy. As the country has allowed energy prices to increase toward international market rates and as shifts occur toward more reliance on service rather than manufacturing sectors, pressure on the environment (including air quality challenges like $SO_2$) can be expected to ease somewhat more.

Still, the established approached for dealing with LRTAP negotiations and deliberations was followed again, and the evidence suggests that those most directly involved have focussed attention on implications of the newer protocol for plans and developments within Hungary. The expectation of significant economic growth in the upcoming years compels decision makers not to treat the future commitments too lightly, and the analytical capacity to project emissions levels and consider options makes the task of domestic discussions easier than was the case during the mid-1980s. (The context for implementation is considerably different, however, as explained later.)

In late 1994 the newer coalition government signed the protocol that had been negotiated largely under the authority of the predecessor government. And since that time, even during the negotiations period, the commitments involved in the second $SO_2$ protocol began to be considered in domestic decision making. Among the actions triggered both by the LRTAP commitments and the decisions of domestic leaders was the initiation of a revolving fund at the level of 30 to 40 million ECUs over the next five years to assist the transitional economy with energy efficiency and energy rationalization goals. In addition, the most recent intersectoral action program has involved the expenditure of 200 million HUF on the development of on-line measuring stations at 48 new sites, in an initiative financed by the Central Environmental Fund. Additional projects aimed at emissions reductions have been initiated, and others are recommended or planned. Anticipated constraints from future protocols as well as other international environmental agreements are increasingly considered as part of the intersectoral program and the use of Central Environmental Fund support.

Beyond such measures, those involved in governmental efforts in certain key ministries like industry have sought to use the ties developed during the earlier state-centered domestic regime to continue informally the kinds of coordination with centralized sectors of the economy. For instance, two major energy-industry firms have dealt directly with the government (in the interest as well of their own needs for technological changes and increased efficiencies) to concert investment plans with changing government emissions requirements. The Hungarian Electric Company (MVM Rt) had been interested in replacing old boilers because of efficiency problems, and the tightening emissions needs of the government fit with the long-range interests of the company. In the case of the largest oil company in the country (MOL Rt), interest in higher quality, desulfurized gasoline has been building. Again, the company's needs were seen to be consistent with developing governmental policy and plans regarding emissions.

In a fashion, therefore, the legacy of the state-socialist style of coordination has assisted in informal adjustment to new international commitments, at least in those sectors in which sufficient concentration economic activity has permitted nonmarket signals to amplify and guide more price-based adjustments. Some have criticized such initiatives indirectly, preferring instead 'transparent subsidy schemes and procedures' on grounds of efficiency and integrity. 'Past habits [have] brought about privately negotiated deals between "powerful" official[s] and company managers', in the words of a recent report (UNECE 1996, paragraph 55). However, at least in a few sectors such informal efforts may continue to be used to concert action toward goals established in international agreements.

## The Second Nitrogen Oxides Protocol

Negotiations continue on a second $NO_x$ protocol, but there is no final agreement and no specific commitment from Hungary in this regard at the international level. The Hungarian process being used for this effort is again similar to that employed for the last dozen years. In the negotiations, Hungary and some other countries have favored taking local and regional effects into account, rather than relying on a European-wide (critical loads) model. The latter position has been favored by the English and Dutch, for instance.

One point worth noting, regardless of the details ultimately enacted, is that the presence of the acidification regime and the likelihood of new commitments has influenced domestic decision making. Aside from other efforts mentioned earlier, several initiatives in the transportation sector have already been introduced; one motivation is to reduce $NO_x$ levels in anticipation of additional restrictions.

Some examples can be mentioned. During the 1990s, the government has offered modest incentives to drivers of older, polluting automobiles to remove these vehicles from the road. (Delivering a Trabant to the government produces in exchange free public transit use for a family for two years.) A second recent program seeks to encourage the retrofitting of existing vehicles with catalytic converters. (The current average age of vehicles operated in the country is 11 years.) And efforts have been made in some urban areas to equip busses with cleaner and newer motors and to improve public transportation. Aside from the limitation imposed by stringent budget constraints, Hungary faces difficulties exercising a great deal of control over transportation because of the heavy levels of international traffic, particularly due to trucks. Some weekend restrictions on trucking are being planned.

This coverage of the Hungarian approach to LRTAP has treated many dimensions of the case and has provided information regarding ultimate results. The pattern of domestic implementation, however, has been dealt with only indirectly and briefly. In particular because of the significant changes in the Hungarian setting in recent years, the implementation question merits further attention. Chapter 11 is devoted to this topic.

# 11 LRTAP implementation within Hungary

The foregoing coverage of LRTAP and the Hungarian experience suggests certain important common features, including: considerable stability in institutional arrangements and processes used for dealing with the acidification regime; increasing competence with and anticipation of the implications of international agreements; a record of success at meeting international commitments, with the substantial 'assistance' of the economic dislocations; evidence, nonetheless, of governmental response to deal with international commitments; and evidence of some impact of CLRTAP protocols on programs, understandings, and emissions levels within the country.

Still, this book has not yet given careful attention to how national decisions and efforts are converted into practice. For some purposes, simply recording the evidence of impact (emissions results) might be sufficient. However, with the multiple influences at work now in Hungary, it can be important to characterize the implementation process in its own right and determine what possibilities and limitations can be identified in the channels between international negotiators and those who seek to solve the practical problems of air quality at the bottom - the 'shop floor', as Hanf puts it (1994a).

Especially given the wrenching changes of recent years, the emergence of more democratic politics, the developing role of local governments and to some extent nongovernmental organizations (NGOs) in environmental debate and decision within contemporary Hungary, and the impact of the market and privatization, it is useful to treat seriously the implementation question - what happens after national decision makers reach consensus about how to deal with acidification commitments at the international level?

The coverage in this chapter provides a general overview, with attention to the principal actors involved, the process used to try to bring national initiatives into practical effect, and some implications of the recent waves of privatization and the emergence of local governments.

### Implementation Challenges

Who is involved in seeking implementation of nationally-established decisions? Once again, the actors centrally involved in the prenegotiations and negotiations, especially at the Institute and the Ministry, are also those most directly involved in developing the plans for implementation.

Indeed, they have been typically directly involved in any training of headquarters or field specialists that may be necessary as a consequence of new targets, approaches or policy instruments. And the interministerial steering group considers on a yearly basis the extent of progress in meeting the objectives of the protocols, as well as the major unanticipated events and potential new lines of effort that may be required if the nation is to adhere to its international obligations under evolving circumstances. The most important evidence of close coordination, then, derives from the identify of the same core actors at the center of these different but closely-related streams of activity.

Some national actions are virtually self-enforcing. An example would be a governmental choice to commit funding at certain levels toward established initiatives. Shifts in incentives or rules, on the other hand, are likely to require either transformation into the nation's legislative framework or incorporation into the regulations and instructions that guide the operations of those in the regional environmental inspectorates.

When the government itself adopts some version of the implementation plan - and there can be delay at this stage because of the policy implications of some of the actions included - further processes are initiated that result in drafting the proposed laws, regulations, instructions, and programs that constitute, or at least authorize, the more permanent effects of the protocol in question and its national approval.

Those close to the center of the LRTAP negotiations and prenegotiations, for instance, who also have been directly involved in developing governmental implementation plans, were involved as well in drafting the 1986 legislative changes in the air pollution abatement provisions of national law. And they have also been active on the proposed new air pollution abatement act, which is planned for introduction in Parliament in the latter part of the 1990s.

The bill exists now in reviewable form but is being held until after passage of the proposed new comprehensive environmental law currently being considered in Parliament. The new air pollution legislation, if passed, is designed among other things to bring Hungarian law into harmony with several of the important European Union standards, including the Regulation on Large Combustion Plants. The draft is based on Austrian and German approaches and will prescribe stricter commitments for industrial and other sectors. The larger companies have already been informed of the likely forthcoming requirements and are seeking - at least in some cases - to adapt to the requirements. (Compliance for small companies may be more of a problem, as is getting information to them in advance of formal parliamentary consideration.) Plans exist, furthermore, to develop a vehicle emissions monitoring system. This issue, under the jurisdiction of the ministry of transport, is receiving attention from this ministry's background institution - which is also coordinating with the Institute for Environmental Management on the matter.

Past the drafting of implementation plans, of course, must come action in the real world of pollution-producing human activities. What is clear is that despite the extensive planning and coordination at the previous stages, implementation experiences gaps in practice. These seem to flow from multiple sources, including: crosscutting pressures from the fragile economic circumstances, inadequate policy instruments, resource restrictions, institutional limitations related in part to the current politico-economic transition, and (what is also an institutional question) a lack of coordination among the several units involved in converting policy intention into results 'at the bottom'.

The treatment of policy in Chapter 6 reviewed the existing situation and offered some coverage of the kinds of gaps extant in the armamentarium available to the government. Similarly, the characterization of organizations and institutional capacity, as explained in Chapters 7 and 8, provided information on the general administrative difficulties under the present system. Only limited additional comments need to be added on this issue. The other factors can be covered in a bit more detail.

First, the difficult economic circumstances of present-day Hungary have tempered in practice what happens during implementation, even as the introduction of market forces holds the potential to encourage efficient use of resources and cleaner technologies over the longer run.

This limiting of implementation happens, in part, even at the prenegotiations stage, as some interests (industry, and the Ministry of Industry and Trade) emphasize the importance of allowing some leeway, and some breathing space, to hard-pressed industrial sectors. It happens also in the budget allocation process for implementation (to be treated shortly). And it happens at the level of regulation in the field, as inspectors - as is done everywhere - exercise discretion in light of circumstances on the ground: good faith efforts on the part of struggling plant managers, local levels of unemployment, differing degrees of seriousness of regulatory offenses, heavy caseloads and the need for selectivity in careful reviews, feasible schedules for eventual implementation, and so forth.

Further, in a point related in turn to the inadequacy of the policy instruments currently employed, the implementers in the field are somewhat limited in their capacities to control violations by the very structure of the fines mechanism, which serves little as a deterrent. Field inspectors, realizing this fact of regulatory life, have even more reason to bargain and seek a *modus vivendi*.

On the issue of resource limitations, little needs to be said. Hungary has been burdened by heavy foreign debt, the government itself is constrained by the bleak economy and widespread tax evasion, and the new coalition felt compelled to enact budget cuts affecting the ministry for environment, along with other policy sectors. All units have felt the squeeze, including the regional environmental inspectorates. The Ministry has drafted proposals to add funding in certain crucial categories. But resource restrictions remain severe. Assistance from abroad for programs directly or indirectly related to air pollution has provided crucial help, as explained earlier, but has also been criticized as poorly targeted. In light of these sobering realities, it should come as no surprise to learn that some implementation problems have emerged because ministries and field units have been overburdened with work or saddled with outmoded administrative or technical support.

Beyond these points, further attention to the institutional setting, especially during these years of transition, can prove instructive in helping to explain performance gaps during implementation. With regard to the implementation of LRTAP protocols, in particular, and improvement in air quality, more generally, two institutional features of the current situation reaching *beyond the national government* can be singled out for particular attention: impacts of privatization and the market economy, and consequences of the changes at the local level. (Other institutional issues are either addressed following these or are covered elsewhere in this study.)

## Impacts of Privatization and the Market Economy

Impacts of privatization and the market, as hypothesized earlier in this study, have been double-edged. Privatization, and the attendant liquidations, affects environmental quality in many ways. And even in terms of direct effects on the execution of national policy for air quality, the proliferation of large numbers of economic actors, many of which are not under direct governmental control, has meant an increase in workload for enforcement staff in the regions. Further, the movement toward the market has provided some incentives toward energy efficiency and a shift of certain production processes and products to enhance competitiveness in international markets. The addition of policy instruments tapping economic incentives in the interest of environmental quality could be particularly influential positive forces; these kinds of initiatives have been widely discussed but largely omitted in practice thus far, as indicated earlier.

Additional, more subtle impacts are also worth noting. In particular, the kinds of coordination possible - and regularly exercised - between national negotiators/ policy makers on transboundary air pollution agreements, on the one hand, and industrial representatives, on the other, face institutional limitations in the current setup. Some of the leverage in the hands of central negotiators has increased under market conditions, as explained earlier. However, what has yet to develop is extensive coordination and representation of the interests of those populating the atomized smaller industrial sectors; in some cases, as well, certain subsectors comprised of several larger plants or firms have also not developed the coordination necessary to allow for negotiations and information sharing in a common, recognized, and accepted forum.

In many subsectors of the economy, there are no respected and well-connected representatives who can advocate the positions of the industry to the central decision makers and, in turn, translate the coordinative efforts of the central actors back to those managing and owning the firms. There is no established forum for this sort of interaction as of yet. In the words of one participant, 'the institutional capacity is not yet there; it is evolving'. Or, as another puts the difficulty, 'we don't have the bridges' to link central initiatives with actors in diffuse sectors of the economy. And NGOs have thus far been relatively little involved as intermediaries in these implementation processes.

In some cases, industrial sectors have formal representation through certain individuals who occupied official roles prior to the politico-economic changes, but these individuals do not command the respect and support needed from those in industry in the boundary-spanning roles. Until central government representatives can develop effective communications links with these parts of the economy, the capacity to deliver on the intentions contained in national commitments to international protocols will be limited.

Some involved in the implementation efforts at the national level look back with nostalgia toward the era of state-socialist governance, when the advantage of centralized directives and financing allowed planners to map connections between broad decisions and at least the formal commitments of several institutions involved in air quality. Today's dizzying pace of change and lack of formal control at the center seems to them disorganized, haphazard, and ultimately ineffective. Whether the contemporary gaps and glitches in the implementation network are a relatively transient property of a transitional regime or a symptom of deeper and more long-lasting incapacities is a matter in dispute among observers and participants. Those who emphasize the potential efficacy of market-based instruments and the importance of developing modern, horizontally-effective public management (see below) view the current phase as painful but ultimately solvable. Those who believe fundamentally in the superiority of one-party systems and their associated planning efforts are profoundly ill at ease in the current period.

## Local Governments

The other aspect worth considering here of institutional capacity and its limitations for implementation during the transitional period has to do with the role of local governments. As explained earlier, these governments are granted independent roles under the current constitutional and legislative framework, and these roles have the potential to influence environmental action generally and air quality in particular. Interesting complications can be seen by comparing the role of local councils in the period before the political changes with the current situation.

First, although the local units have been granted significant roles, in many municipalities the personnel involved are the same as in 'the old days'. There are complaints about the continuing legacy of the bureaucratic attitudes familiar from the days when local units were primarily instruments of the central party apparatus rather than locally-responsive structures populated by individuals exhibiting a public service ethic and sense of democratic values. In part, therefore, this story is one of old styles of operation dying hard in the wake of political transformation.

Second, in almost all the local units, there is insufficient expertise available to create the capacity for effective action. Budapest and a handful of other cities, mostly of medium size, are currently capable of addressing some environmental questions, and a few are beginning to do so. However, the vast majority of local units are exceedingly small and also inexpert. Environmental questions were not directly addressed before the political changes, the local units have few capable experts currently on hand, and they are so hard pressed from a number of directions - including financially and in terms of economic distress among local populations - that they are taking relatively few initiatives in this direction as of yet.

Third, the county level of government (located between the central government and the municipalities), which might eventually be the locus of more region-wide environmental expertise, is at present weak in general and almost completely without jurisdiction on environmental questions.

Fourth, thus far localities have only just begun the process of networking among themselves in horizontal fashion to approach regional questions in a coordinated fashion, pool expertise, and share experience. Regional associations of local government have begun to form and serve as forums for information exchange and occasional training sessions, but these developments are sporadic and tentative. Even in Budapest, a city in which many local functions are in the hands of the 22 smaller 'municipalities' comprising the city proper, relatively little coordination has developed on a whole range of policy issues.

Fifth, even the basic legal questions regarding the authority of local governments to become directly involved in controlling and enhancing environmental quality have been resolved only slowly.

Some observers and participants in the air quality implementation institutions believe that the political and economic changes may have even limited the possibilities for local influence - thus 'squeezing' these governments between their responsibilities, as suggested in political rhetoric and the expectations of their citizenry, on the one hand, and their actual ability to influence air quality, on the other.

The reasoning here derives from knowledge of how local councils under the old regime sometimes worked in practice - in contravention of official rules. Before the political changes away from the one-party system of political rule, bargaining was possible between officials who were part of the local council and managers of industrial facilities within the territory. Local officials could sometimes use informal contacts as leverage for pollution abatement. Sometimes, as well, the local officials would exercise their leverage in other directions: for instance, by suggesting that they would tolerate levels of emission that were officially in violation of standards in exchange for 'contributions' from industry in support of a local school or park that had not received sufficient allocations from the state apparatus.

Since the political changes, these channels of leverage, oiled by party ties within the state-socialist apparatus, have weakened. Where polluting industry is in private hands, the official national policy of emissions standards remains; but of course, as indicated earlier in this analysis, that policy provides inadequate incentive in most cases to cause compliance merely via financial inducement. The informal points of leverage have been broken by the collapse of the state-socialist apparatus. And in this respect, municipal officials may be in a less influential position regarding certain important regulatory targets.

One knowledgeable and experienced participant in the air pollution control system in Hungary characterized the present situation as regards local governmental involvement as one of incomplete institutionalization. Ideally, it is suggested, local involvement would be as one corner of a triangle consisting of industry and the regional environmental authorities as well. With sufficient experience and institutionalization of coordinative processes, the three kinds of parties might be able to work effectively together toward acceptable options. However, it is offered, the current state of affairs more closely resembles binary relations between regional inspectorates and industry, with localities thus far not involved in regular, stable forums of environmental discussion and decision.

**Limits to Coordination**

One additional limitation on policy implementation deserves attention in the present context: the lack of coordination among even official, formally-involved institutions at or near the 'bottom' of the implementation chain. As mentioned above, this topic has mostly been treated earlier in the review of the formal institutional setting for air pollution policy execution in Hungary. However, an additional issue can be raised here, one especially apropos regarding connections between or among decisions. The matter is the managerial perspective that is largely institutionalized in the Hungarian administrative apparatus - a public management approach that experiences additional problems as well stemming from the institutional fragmentation of the formal apparatus. An explanation is in order.

A longer-term legacy of the Hungarian experience is the approach to public management that is still visible in the implementation of air pollution policy, as well as in other fields. Hungary is a country in which a legalistic tradition has long guided a good deal of what happens in and through public bureaucracy. There has been little borrowing from the ideas developed in contemporary fields of public management and policy implementation, little looking beyond the profession of law and the drafting of regulations to energize public action.

Further affecting the Hungarian approach to implementation, of course, the decades of state-socialist rule have had their own impact: insulating Hungary from the impact of modern management ideas in government, limiting the development of a public service ethos, greatly restricting the tendency of public managers to develop service-oriented administrative units or channels of popular information exchange, and preventing the development of experience with such important features of implementation in mixed-economic settings as budgeting and accounting. Perhaps most of all, the notion of an active specialty and process of management, apart from the successive refinement of regulations and standards, has been lacking. How to connect the parts of the mechanism, especially for elements that have acquired considerable independence, even formal autonomy, is a question that had not been much asked in recent decades.

The practical implementation tasks of coordinating diverse but interdependent participants toward programs exhibiting some degree of common purpose across different institutions and interests - these were not a focus of the state socialist period. How to take advantage of price signals, incentive-based tools, and individual and organizational motivation toward cooperative effort while also utilizing the unique capabilities and perspectives of diverse ministerial and societal actors - these were not issues on the agenda of line managers.

Now, with the multiple trends of institutional change visible and consequential - from privatization and market development, to the emergence of interested and increasingly knowledgeable NGOs, to substantial local autonomy and the need for development of intergovernmental ties for programmatic ends, to the dependence of successful policy execution on interagency cooperation at the national level and in the field units of ultimately-national authority - there is a pressing need for implementation managers who are skilled at: establishing and utilizing forums for institutional cooperation and coordination; taking a broad view of their responsibilities to include developing links across organizational units, encouraging communication channels across multiple levels of formally separate governmental units, and actively shaping and stimulating interunit agreements and common understandings and quid pro quos toward the end of more effective program execution; and simply bargaining and exerting diplomatic skill to move the multiple institutions and interdependent actors toward the desired end.

During the state-socialist era, not only were these kinds of behaviors not encouraged or rewarded, but the very notion of informal communication and coordination horizontally threatened to undermine the source of state and party authority. It was systematically discouraged. The informal ties that mattered were less those based on task interdependency and more on personal and political links developed and nurtured through the system. Differently-situated professional specialists developing task-based links toward the end of improved program effectiveness were not an important part of the picture.

The style of 'management' prevailing for years, then, was one with largely passive program administration, little systematic use of information systems or modern notions of human resources development and training, and even less reliance on public service motivation or horizontal task-related ties.

Examples of the continuing legacy of these earlier styles of management are visible today in Hungarian implementation of air quality efforts. For instance, in the process of executing new initiatives that may be technically required by international (LRTAP) protocols, have been considered for the development of governmental directives, and have been fleshed out somewhat in an implementation action plan, revised regulations may be drafted. How are these translated into the language and routines and priorities of the field officials in the regional environmental inspectorates? The answer is that the central ministry for environment makes the new regulations available to field officials via the government publication containing such new formal decisions. At that point the central Ministry has completed its responsibilities and exercises no oversight of the inspectorates. Headquarters efforts are devoted more to developing the periodically-required reports to the United Nations Economic Commission for Europe on data from the monitoring system - this task too done in consultation with the background unit, the Institute for Environmental Management.

Those in the field, in turn, are expected to know of such regulatory changes. But the responsibility of making the connection and ensuring a revised program of practical action is left somewhat vague. Ensuring compliance is limited to responding to questions from the field. As elsewhere in this society in transition, the meso-level coordinative mechanisms are wanting.

For certain kinds of initiatives or changes, the ministerial staff may offer comments and advice at the annual national meeting of inspectorate personnel involved in air pollution abatement. Important new regulations are likely to receive some attention in this forum. If a new rule explicitly also requires training of field personnel, the Ministry also organizes this activity. In fact, this sort of effort as well has typically been executed from the office of the Institute of Environmental Management which had been so centrally involved in the other phases. But overall, the system is noteworthy for the relative lack of active implementation links across levels and institutions at these phases of the process.

Further, the impediments outlined here appear to be exacerbated by the almost complete absence of formal horizontal links in the administrative arrangement for air pollution that was sketched in Chapter 7.

All in all, the conversion of reduction plans and implementation schemes into cooperative, effective efforts toward improved air quality is the largest question mark in the Hungarian system.

This is not to say that the system never works, for Chapters 9 and 10 provide evidence to the contrary. But it is important to note that most informed observers of the Hungarian efforts currently underway see a significant problem in the 'action' phases of the overall effort, and some are cynical about the continuing prospects for converting protocols into genuine government commitments backed by needed budgets and across-the-government agreement, particularly when it comes to apparent tradeoffs with the economy.

Despite the striking continuity between the negotiating team for LRTAP and the national actors responsible for developing air pollution policy, and despite the degree of stability at the core of the negotiating effort, a number of those involved in environmental policy and air pollution control are critical of the efforts made thus far in concerting action through government to achieve policy goals, whether internationally-driven or otherwise. In the near future, therefore, Hungarian success in its participation in the acidification regime may depend heavily on progress in the implementation sphere, as well as on the development of more effective policy instruments suitable for the nation's altered political and economic circumstances.

In the next chapter, and in light of the detailed information presented in the last three, an overall review of the impact of LRTAP on the policy situation in Hungary is sketched - as is the reciprocal channel: the ways in which the national experience, particularly during and since the first sulfur dioxide protocol negotiations, may have influenced Hungary's approach to and involvement in later international negotiations and cooperative possibilities.

# 12 LRTAP and Hungary: overall influences

What can be said regarding the overall impact of the Convention on Long-Range Transboundary Air Pollution in Hungary, as well as the influence of national experience on participation in the LRTAP regime?

While many political and economic changes have taken place within the country since the initial LRTAP negotiations, the following generalizations summarize a good deal of the mutual influence of national policy action (and its implementation), on the one hand, and the international regime, on the other:

- National policy has not been reshaped in fundamental ways by the international regime. Instead, modest initiatives and adjustments involving policy-oriented learning have characterized the Hungarian response (Sabatier and Jenkins-Smith 1993);
- Implementation has been influenced, but the selection of policy instruments and weaknesses in the institutional arrangements for program execution have limited the effectiveness of national response;
- LRTAP agreements have had ramifications for institutional adaptation, policy process, government program, and emissions levels. However, impacts on emissions have been less significant than have both the economic decline and decisions reached on other grounds;
- Hungary's experience with and following participation in international environmental regimes has had subtle impacts on the nation's approach to renegotiations for LRTAP; and
- The continuing development and elaboration of the acidification regime has encouraged an increased attention within Hungary toward the nation's international commitments and obligations. The recent domestic difficulties, particularly with respect to eco-

nomic matters, have limited but not entirely negated the interest in the acidification regime.

Several of these points have been documented in detail in preceding chapters. Additional comments and observations follow.

## The Shaping of National Policy and Implementation

The clearest evidence on lack of fundamental impact by LRTAP and its protocols on national policy in Hungary is the fact that the basic policy containing guidance on air pollution issues remains the 1976 law, obviously in effect prior to LRTAP. There have been modifications in national policy, as mentioned earlier, but these were not influenced heavily by the protocols or by LRTAP overall (see Bándi, Faragó, and Lakos 1994).

Instead, policy ramifications have been more subtle - evident in low-visibility initiatives and in the difficult-to-document influence of international agendas on the muted debates of interministerial committee and background institute discussions. Those close to the core of the domestic acidification effort have used ingenuity and initiative to make incremental adjustments in an overall system largely preoccupied with other issues and incapable of comprehensive reform on the schedule driving decisions at the international level.

The process used in Hungary during the last several protocol post-negotiation periods has involved a conscious effort to connect commitments made in the international forum to Hungarian strategies for air pollution control, Hungarian directives, and nationally-established implementation plans. All evidence suggests serious attention to the issues at stake and intentions to find ways of ensuring compliance. Results have been documented in earlier chapters and show both programmatic adjustments, such as the several initiatives in the transit sector, and informal coordination for implementation, like encouraging investment in cleaner technology for electricity generation in anticipation of protocol restrictions under negotiation.

Efforts to trace the influence of acidification protocols on events in Hungary, furthermore, particularly in the context of general environmental policy stability prevailing in the country, inevitably encounter ambiguous and complex evidence. An instance of this type, not mentioned above but occurring at the time of the political changes at the end of 1989, is a program established to reduce emissions from the use of fuels; the effort was undertaken with the active support of Germany (and would likely never have happened without Germany's assistance). The 'coal aid' involved Germany's donation of a considerable quantity of 'free' coal as assistance prompted by the shift in political regimes.

Germany requested no payment for the coal but did stipulate as a condition that the equivalent market price for the coal be placed by Hungary into a special account. Access to these funds could be obtained by industrial units or organizations interested in initiating technological changes within their sector or company, if it could be shown that these changes actually trimmed the level of emissions due to the use of coal, oil, or other fuel. Hungarian studies suggest that this program itself did have an impact with regard to $SO_2$ emissions, and that the funding did not simply subsidize what was happening anyway. The program encouraged some technological modifications that would not have been taken, or at least would not have been taken at that time.

Some involved in this effort indicate that although this program was not technically derived from LRTAP triggering protocols, it was accepted more readily in the first transitional government because of international agreements and the perceived need for Hungary to commit (or recommit) itself to the acidification regime and other international environmental responsibilities after the political changes. The program offered by Germany provided an opportunity to link some concrete actions with the stated governmental commitment. Under the circumstances environmental advocates, and those sensitive to national commitments in the European region, were provided leverage that they might not have had in the absence of LRTAP.

The coal aid example illustrates the limitations inherent in any effort to document precisely the influence of the international acidification regime on domestic activities, since it suggests much more complex causal patterns than the simple unidirectional, top-down image often considered in modelling the impact of international agreements. The possibility of indirect, longer-term impact cannot be quickly dismissed.

Still, the main tenets of the Hungarian approach to air quality in general have not changed in a generation, and no international protocol has been a sufficient stimulus to shift the focus away from heavy reliance on an after-the-fact emissions control supported by a fine-oriented system that only mildly penalizes violators. The adjustments prompted by the acidification regime have been low key and have always been woven quietly into the details of programmatic effort by the relatively few central actors at the core of the effort. Even in other units of the environmental ministry, little is known about LRTAP and its implications. And the same can be said, more strongly, regarding the larger institutional matrix that forms the core of environmental program execution in Hungary: the other ministries, the regional inspectorates, and the local governments, not to mention the NGO community, the political parties, and the larger public. LRTAP and its protocols do not play a conscious role in the day-to-day choices of most actors involved in controlling air pollution in the nation.

And the requirements of the international regime have continued to be addressed through an implementation system with notable weaknesses - including the inspectorate system, the separation of significant air pollution responsibilities into several ministries and administrative units, and the relative neglect of integrative mechanisms. The lack of institutional capacity available in such a system is highlighted under current conditions, with increasing numbers of regulatory targets, massive shifts in property, increasing market dynamics, enhancement of local responsibilities, and the opening up of public life to the requirements of democratic access.

## Impacts of LRTAP

Overall, the reductions in sulfur dioxide, nitrogen oxides, and VOC emissions during the last several years can be traced primarily to the severe economic difficulties in Hungary. Some cuts have been due, as well, to decisions consciously made by government but not as part of a strategy of environmental quality or compliance with international commitments. The ultimate impacts of the CLRTAP in Hungary have been much less grandiose than some proponents might claim.

But the agreements have not been completely impotent. The acidification regime has triggered the institution of a domestic institutional center for acidification efforts, initiated a process of domestic discussions and deliberations, encouraged the development of analytical expertise, and resulted in identifiable programs and even emissions impacts. These last results are impossible to pinpoint in precise quantitative terms, but the evidence gathered in this investigation indicates that they are probably of modest but not insignificant proportions. Furthermore, as the next chapter indicates, LRTAP along with additional encouragements toward international collaboration on environmental questions may be encouraging the development of a perspective within Hungary that may have significant influence on domestic efforts over the longer term.

## Feedback between National Policy and Implementation Experience and Continuing Participation at the International Level

It is obviously too early to draw firm conclusions about the ways in which national experience under LRTAP influences Hungarian renegotiations and continued participation in the international regime. Therefore, the following comments must be regarded as preliminary and tentative.

First, the evidence available suggests that the continuing participation by Hungary in LRTAP and the increased understanding of the acidification regime and its implications have drawn Hungary into steadily closer and more serious treatment of the international dimensions of its domestic decisions. The nation is now using its leverage as a transitional setting to bargain for more leeway and more time in dealing with the difficult implications of future commitments; but the international expectations and intended influence of the acidification regime is equally obvious among those at the core of domestic decision making on air quality. The acidification regime, supplemented by international aid and the complementary support available from other international agreements, is likely to exert leverage into the future.

And second, some subtle impacts of participation in the acidification regime can already be identified. One internationally-experienced environmental official in Hungary indicates that international bargaining has been a process of education and socialization that has influenced strategies, tactics, and communication channels both domestically and internationally on the part of the negotiators in question. Also, the sheer number and complexity of international environmental agreements in which the nation now participates - including those outside the sphere of acidification, narrowly-construed - have increased the awareness within government of the role and relevance of international commitments in domestic decision making and has encouraged 'networking' among the relevant domestic policy makers and advisers, as well as within the relatively small group of delegates to the various sets of international negotiations (see the following chapter). These maturation effects are not surprising, given the multiple functions that international regimes like CLRTAP can perform (for instance, as sketched in Helm and Sprinz 1995).

Indeed, the degree of participation by Hungary in cooperative efforts encouraged by the acidification regime have gone considerably beyond good faith efforts at domestic compliance. A recent summary of the extent of participation by 39 countries in programs involving data collection and task force initiatives to support the acidification regime indicates that Hungary has become more heavily involved than most countries, with a range of efforts placing among the top third of the European nations involved (see Economic Commission for Europe, 1995, p.136).

This chapter places the LRTAP experience in broad perspective. Before this investigation is concluded, nonetheless, it is helpful to consider one additional dimension of international influence over Hungary's air quality efforts to deal with acidification issues: the emerging importance of the European Union. This subject is impossible to address completely, since Hungary's set of adjustments to the EU are in flux and the nation is not yet even a full member of the Union. Preliminary observations are the subject of Chapter 13.

# 13 Hungary and the European Union

The influence of LRTAP on developments in Hungary in the future must be seen in conjunction with what will soon be a much more consequential channel of international impact on environmental matters: the European Union.

Hungary currently has associate member status in the European Union and is expected to be accepted as a full member soon. The timetable has been uncertain, but the intention has been firm in both of the two domestic government coalitions since the political changes of 1989. Support for EU membership crosses virtually all political parties with any popular following within Hungary.

## The EU and Harmonization

As part of Hungary's increasingly European outlook in national policy, some institutional changes have been made within government. And a number of efforts have begun with the purpose of 'harmonization': altering Hungarian policy, particularly formal law, so that national legislation comports with the requirements of the EU. A portion of the harmonization effort is being devoted to environmental policy generally, including acidification issues in particular. The general commitment to harmonization is now expressed in broad national policy concepts on the environment (Ministry for Environment and Regional Policy, 1995, p.39), and more specific harmonization efforts are underway.

Both of these shifts - institutional and policy-focussed - should be elaborated upon. First, however, an obvious point should be made to establish context. Until the last few years EU membership has not been an option that could be considered openly on the domestic political agenda. Institutional and policy development toward harmonization was completely absent until after the political changes. Thus, for instance, influence within Hungary of the EU regulations on large-combustion plants or vehicle emissions has been virtually nil until very recently.

It is true that in some quarters Hungary was seen as more friendly to Western Europe and EU policy orientations on the environment than some of the neighboring states. But Hungary was not part of the international regime being guided by EU regulations, did not participate in their debate and development, and operated its own national policy-making apparatus divorced from the constraints and decisions of the EU - certainly on such issues as emissions causing air pollution.

Nevertheless, since the political changes movements in the direction of the EU have been notable at the national level. The first post-socialist coalition established a Ministry for International Economic Relations, in which was placed central responsibility for coordinating the harmonization process in Hungary. The new coalition has eliminated the Ministry but retained the function, which has been moved to a unit within the Ministry of Industry and Trade. This office on European Integration Affairs is charged with EU coordination. Its function has become steadily more important in the recent period.

On issues relevant to air pollution and the regulations of the European Union, the office works with the air protection department of the Ministry for Environment and Regional Policy. Also likely to be important in the future are the national offices opened in Brussels to liaise with the EU.

Some participants and observers in the environmental policy field in Hungary are nonetheless skeptical of the importance of environmental issues in the generally well-accepted Hungarian commitment to harmonization and EU membership. Some see governmental discussions of harmonization as focussed almost exclusively on economic, especially 'market', considerations. Opening the markets, rather than regulating or controlling economic activities within generally-market settings, is what the government most consistently emphasizes. Therefore, some participants worry about the ultimate fate of environmental considerations in the policy and implementation efforts to be undertaken in the coming period.

In particular, analysts note two significant difficulties that are likely to constrain EU influence over environmental developments in Hungary:

lack of integration of environmental policy into other related sectors, and weaknesses in implementation (see, for instance, Bándi 1993). Regarding the first point, 'There are some indicators of integrative thinking in the energy-environmental area. . . . However, the role of environmental concerns is generally considered to carry very little independent weight in the political process. . . . [A]lthough Hungary has signed a number of international environmental agreements in the past, in only one case has it enacted implementing legislation' (Matláry, 1994, p.149).

And this observation, in turn, is connected to the implementation issue. As Matláry points out, and as is consistent with the coverage earlier in this study, 'Hungarian environmental policy . . . exists, but its implementation is regarded as highly ineffective' (1994, p.148). So realism and the experience of recent years should constrain reasonable expectations.

Nonetheless, Matláry's balanced conclusions are apropos:

> The potential role of the EU in forging an energy-environmental policy in Hungary is . . . much greater than its present role. . . . On the Hungarian side we may assume that the desire to join the EU will increase the political willingness to adapt to its environmental policy as well as to create conditions for effective implementation. Implementation will probably remain the major problem, as it is in many EU countries in this area, and the political process of integrating environmental criteria in other policy areas can also be assumed to present many future political conflicts. However, in both these regards Hungary is [in] a position no different from that of other European countries. [1994, p.151]

Initiatives have been underway to influence harmonization on environmental issues, especially air pollution and acidification. One effort of importance is an initiative funded by PHARE. Half of this sizable project is being used for harmonization of environmental law, the other half for training of environmental personnel regarding EU requirements and the importance of national compliance. This action is likely to have a significant impact on environmental law and its implementation in Hungary, and certainly the assistance can be expected to have a measurable effect on the institutional capacity of Hungarian environmental policy management.

The ministry for environment has also contracted for studies on the subject with an environmental consulting firm. The results of these preliminary efforts have served as background. Similarly, the independent Environmental Management and Law Association (EMLA), a private 'public interest' organization, has developed its own review of requisites for harmonization in the environmental field, including on the subject of air pollution policy. And NGOs have begun to attend to the harmonization effort as well.

The core of the air pollution policy 'influentials' have already drafted a revised national law on the subject, in part with a view toward the harmonization objective. And its requirements are already affecting the informal communications developing between governmental officials, particularly in the ministry for environment (sometimes aided by information and contacts provided through the Ministry of Industry and Trade), and large industrial sectors over the statutory constraints likely to be in force in the future. Some industrial firms, then, including certain of the largest ones, are even now adjusting to EU regulatory constraints through this somewhat indirect route.

Yet the whole issue is scarcely visible to those outside the immediately-relevant circle of actors. International acidification agreements of the EU, based largely around specific emission targets rather than chemicals, are beginning to have an effect. But the impact is likely to be felt most directly on emissions themselves - in the increasing participation in the Hungarian economy of domestic firms anticipating the introduction of European standards or multinational companies already experienced in dealing with EU requirements - rather than on official national policy. For the legislative changes have been delayed until after disposition of the general environmental bill, and the harmonizing forces of the European market are moving in advance of the formal stipulations of Hungarian policy.

One additional acidification-related initiative is deserving of mention here: an effort being developed through the ministry of transport and its background institute to connect Hungarian policy on vehicle emissions with the EU standards. This process has been coordinated with the revised air pollution legislation developed via the Institute for Environmental Management and the environmental ministry - once again, via informal coordination among the insiders, and through small projects involving contracted work between the two institutes.

Overall, a number of signs indicate that the governmental focus on EU requirements, including environmental standards affecting air pollution, will occupy increasing portions of Hungarian policy makers' limited attention and financial resources in the next few years. The institutional moves that have taken place and the likely budgetary and policy making steps to follow during the 1990s suggest a growing importance for the EU 'connection', with some impact on air quality over the longer run.

Hungarian air pollution policy specialists - with their long experience dealing with international acidification agreements and their fit or link with national policy - are at work with the conscious intent of modifying Hungarian policy and practice to comport with the requisites of this additional, newly-emerging, environmental regime commitment. Eventually, this line of influence, along with other international environmental agreements, can be expected to have more impact than the acidification regime itself - even on those chemicals targeted for cutbacks through LRTAP.

## Additional Considerations: International Environmental Agreements and the Institutionalization of a Transboundary Consciousness

An additional dimension to the issue of control of transboundary air pollution can be brought into the discussion at this point. This book has focussed on the role of LRTAP and the European Union as the main components of international acidification regimes, and the inquiry conducted thus far has explored the extent to which the regimes - defined in this fashion - have had impacts within Hungary, at multiple levels of operation, from national policy making to implementation in the field and ultimately to the goal of reduced (or frozen) emissions of regulated pollutants. The study has also explored some aspects of the chain of influence in reverse, to search for evidence of the impact of the regime and Hungary's participation in it over time on renegotiations, or on operations developing after experience with the international agreements.

These topics help to develop the broad picture of how international environmental agreements on acidification might matter in Hungary. However, when treated in relative narrow terms, as this analysis has done, these considerations are likely to omit another aspect of influence on the Hungarian situation. For LRTAP and the EU are consequential, but they themselves are only part of the full panoply of international environmental agreements and regimes that operate now with some impact on the actions of Hungarian government and, presumably, on the results in terms of emissions and transboundary control.

More specifically, one of the more subtle features of the current Hungarian setting for air pollution control and international agreements is the sheer proliferation of agreements, including some beyond the formal realm of transboundary controls. International agreements on ozone (thus CFCs), the recent climate change convention (and therefore $CO_2$), and other emerging international regimes aimed at affecting the environment, especially air quality and the impact of airborne pollutants, touch upon some issues that are also of concern within the framework of LRTAP and the EU.

Thus, for instance, the recognition that technology transfer issues negotiated internationally can have impacts on the ability of the nation to succeed in its commitments to international environmental agreements has penetrated the consciousnesses of the responsible officials. And the knowledge, to take another illustration, that certain emission sources or industrial sectors might pertain to two or more international agreements has meant that in the few years national policy makers in Hungary have begun to realize that some coordination across ostensibly-disparate national decision settings and regulatory units may be useful. The climate convention, for instance, is thought by some experts closely involved with the issue to pose considerably more severe restrictions on energy use in Hungary than the renegotiated $SO_2$ protocol of LRTAP.

The theme here is the potential for increased perceived interdependence not only of Hungarian moves at the international level with national policy making, regulatory elaboration, and actual implementation, but also 'horizontally' at the international level, across agreements and conventions and protocols.

In Hungary at the present time the behavioral impact of this growing consciousness of international commitments, taken as a set, is just beginning to be observable, and only at certain levels. Those environmental officials with responsibility for developing national policy making regarding some portion of the nation's international commitments have realized that they do not operate in a vacuum, and that ideally they must coordinate their strategies with those being pursued by others, who are also expected to concert national policy with international agreements in related but distinct forums. The interdependence, in other words, extends 'to the side' within the domestic scene, and not simply vertically.

The tightly-integrated horizontal coordination that this recognition might call for has not emerged. The major actors know about each other, and about the 'other' international agreements outside their own purview. But the mutual recognition has not yet crystallized into any formal linkage arrangements, nor into regular channels of coordination and information sharing. Informally, some of the connections are beginning to develop. But no official and no office possesses full information on the complete array of relevant actors, agreements, and national commitments, nor on the technical information most relevant to concerting the agreements in practice.

The dynamics in the near future are difficult to predict. The increasingly shared consciousness on the part of the different governmental officials charged with addressing some part of Hungary's international environmental commitments could well strengthen the collective leverage and visibility of these perspectives and institutional niches in domestic policy making. A 'critical mass' of internationally-aware and -responsible environmental officials (and others) may be emerging, with practical impact on national policy making and implementation. One possible coordination point for such a process might be the relatively new global environmental office of the Ministry for Environment and Regional Policy. This scenario, then, would suggest increasing prominence, and perhaps collaborative learning and more effective influence, on the part of those treating seriously the nation's international commitments.

On the other hand, as can happen in bureaucratic settings anywhere, the proliferation of agreements - resulting in a multiplication of administrative centers of activity - can generate infighting and competition for authority, prominence, and lead responsibility. This tendency may even become pronounced in a setting like contemporary Hungary, where budgets are unusually tight and institutional capacity strained. This kind of dynamic, then, might result in an apparent negative-sum game, with different units and officials canceling out some portion of each others' clout in environmental and broader arenas of decision.

Either dynamic, or some combination of both, would be likely to alter the current loci of influence on Hungarian LRTAP and EU decisions. But the manner of change, and toward what outcome, is less clear. The point, then, is that the recently emerging institutionalization of what can be considered a 'transboundary consciousness' holds the potential to have additional impacts on the actions of the Hungarian government, as well as others within the country, and possibly even on the relative importance of international environmental agreements in Hungary.

# 14 Conclusion

As the preceding chapter indicated, the movement toward European integration has now involved Hungary, and it is likely that the future will bring a shift of attention and participation toward EU issues, policies, and institutions. This hypothesized influence on Hungarian acidification efforts, mentioned in Chapter 3 of this study, is only just emerging but can be expected to shape domestic efforts in important respects.

Other hypothesized influences on Hungarian acidification efforts have also been considered by implication in this investigation, and some of the principal conclusions regarding each can be sketched here.

## Domestic Political Changes

The domestic political changes in the country have affected the context in which acidification issues are considered, and it is likely that as the Hungarian transition continues a more open political system, with more coherent political groupings and the emergence of a stronger environmental voice from NGOs and political parties, will bring changes of consequence to the way that environmental issues like acidification are handled. Thus far, however, the most notable feature of the Hungarian system is how little the altered political landscape has influenced the way that acidification issues are handled: shifts in 'fundamental' influences have occurred, but the Hungarian response has been to try to make adjustments via incremental shifts in 'proximate' factors (Jacobson and Brown 1997). It could be argued that the lack of change itself reflects the broader sense among the public that economic considerations and social welfare concerns should take precedence over environmental questions during this period. From this perspective, the demand-supply politics model of acidification decision making fits the now more democratic Hungarian setting, as it does other countries that have been studied as part of the acidification regime. Careful tests must await shifts in salience of domestic issues as the Hungarian political transition continues.

## Property Shifts and Market Impact

The property shifts and emergence of market forces in Hungary show signs of the double-edged impacts on acidification efforts hypothesized at the outset of this volume. The most obviously heightened implementation challenges derive from the proliferation of regulatory targets and lack of 'bridges' in the increasingly differentiated setting, as well as the slowly emerging involvement of local governments. And the very deliberate pace so evident in the introduction of incentive-based policy instruments is due in part to the 'market as prison' aspect of the emerging mixed economy - as policy makers in Hungary seek to avoid further economic pain by keeping marginal enterprises in business. But market forces also show signs of aiding in the acidification challenge - by discouraging waste, for instance, even if the increased costs of goods like energy impose social distress; and by providing entre to efficient multinational firms - even if some of these seek to apply in Hungary technologies deemed unacceptable in some other national settings.

The ultimate impact of the economic changes on acidification in Hungary, nevertheless, must await a period of stable economic growth and also the introduction and execution of more incentive-based policy instruments.

## The Acidification Regime

In the midst of this series of dynamic shifts, what can be concluded about the independent impact of the acidification regime itself on events in Hungary? As indicated in Chapter 12, the evidence examined in this study indicates that LRTAP has achieved some influence, as measured in a variety of ways both direct and subtle.

In their scholarship aimed at explaining the degree of effectiveness achieved by international environmental agreements, Haas, Keohane and

Levy (1993) point to three requisites: governmental concern, a hospitable contractual environment, and political and administrative capacity. These elements, they argue, support - respectively - three components of effective international regimes: development of supportive national positions, productive engagement in international bargaining, and domestic implementation of commitments.

In the case of Hungary and the CLRTAP, the achievements and limitations of acidification control in turn can be linked to these same components. National concern has been narrowly held among policy elites and specialists, and it has been domestically focussed rather than tied tightly to transboundary impacts. But it has been steady and serious - and buttressed as well even through domestic political transformations by officials serious about delivering on internationally-negotiated commitments. National positions have been, accordingly, technically narrow and incrementally drawn - but supportive.

The evidence is clear, as well, that the LRTAP regime as a forum for cooperation and negotiations has been a political context with alliances and issues aplenty, despite the sometimes technocratic language of acidification discourse. But that is not to say that LRTAP has provided an inhospitable context for bargaining. The evidence suggests that Hungarian participants have been actively and fully involved, and respectful of the acidification regime, even if knowledgeable and protective of their domestic interests and perspective.

Domestic political and administrative capacity have been limited. During the state-socialist period, commitment at the top and political ties to the East were the main stimuli for participation and ensured real attention to the negotiations and agreements developed from international participation. And support for LRTAP has not wavered, despite major changes in domestic public life and shifts in governing coalitions. The core participants in the acidification control effort have been stable, supportive, and knowledgeable. But the institutional capacity to effectuate policy commitments has been constrained - both before the political changes and afterward. A more open political system can sometimes leverage broader participation on behalf of salient policy objectives, but the transitional regime has not thus far been very open to such initiatives. The horizontal and networked relations that may be necessary for managing programs in a mixed economy and complex policy setting have yet to develop. And the process of privatization and introduction of markets have complicated and tested implementation efforts, even as they have created the potential for a more environmentally sensitive political economy in the future.

It would appear, therefore, that domestic capacity has been the principal limiting factor in the influence of the acidification regime in Hungary. It is tempting to suggest, therefore, the importance of heightened attention to this issue for enhancing the future effectiveness of LRTAP within the country. Indeed, due regard for strengthening domestic political and administrative capacity in the broadest sense are matters of high priority, and this study is hardly the first to point to such issues as central.

As the preceding chapter indicates, however, neither acidification nor other environmental questions can be treated in isolation. As Hungary becomes committed to not simply one but a range of consequential international understandings on the environment, the challenge may be to develop capacity to integrate, synthesize, and execute across an array of programs and constraints - opening governmental processes and extending involvement, improving performance while also respecting the limitations inherent in liberal regimes.

This requirement in turn suggests that while administrative progress and budget enhancements can be helpful, no mere bureaucratic fine tuning by itself will effect the political and societal transformations that are required. Achieving environmental objectives in the Hungary of the future is likely to demand both innovative new policies and a transformation in the conduct of public life.

# Persons interviewed

Prof. Dr Gyula Bándi, Member of Board of Directors, Environmental Management and Law Association; Scientific Director, Copernicus Environmental Law Program in the Danube Region, Eötvös Lóránd University Faculty of Law.

Gyorgy J. Bíró, Head of the Department of the Environment, City of Vác.

Prof. Dr Miklós Bulla, Head of Department of Environmental Engineering, Széchenyi István College, Győr; and Director, Institute for Economic and Environmental Development in Central and Eastern Europe (a private institute), Budapest.

Ferenc Burger, Economist, Economics Department, Ministry for Environmental and Regional Policy.

Vilmos Civin, Environmental Manager and Head, Department of Environmental Protection, Hungarian Electricity Company.

Istvan Csallagovits, Department Leader, Sectoral Department responsible for negotiating among ministries, Ministry of Environment and Regional Policy.

Katalin Csorba, Head of PHARE office, Ministry for Environment and Regional Policy; PHARE coordinator.

Éva Fáber, Strategy Department, Ministry of Environment and Regional Policy.

Dr Tibor Faragó, Head, Department of Strategy; former counsel, Bureau for Sustainable Development, Ministry for Environment and Regional Policy.

Zsuzsa Foltányi, Leader, Energy Club and Ökotárs Foundation.

Sándor Fülöp, Executive Director, Environmental Management and Law Association, Hungary.

László Gajzágó, Head Adviser (retired), Air, Water and Soil Quality Protection, Ministry for Environment and Regional Policy; delegate to UNECE CLRTAP until 1995.

Mihály Gulyás, Head of Section for Economic Regulation, Ministry for Environment and Regional Policy.

Dr Iván Gyulai, director, Institute for Sustainable Development, NGO/scientific organization.

László Karas, Project Manager, Environmental Management Training Centers Network, The Regional Environmental Center for Central and Eastern Europe.

Miklós Koloszár, Chief Counsellor, Department of Environment, Local Government and Regional Policy Department, Ministry of Finance; member, Interministerial Steering Group.

Dr Endre Kovács, Head of Department of Environmental Consulting and Engineering, Institute for Environmental Management (Institute for Environmental Protection); member, Interministerial Working Group; delegate (since 1984) at the UNECE CLRTAP; president of Working Group on Technology, UNECE CLRTAP, until 1993.

Zoltán Lontay, Thermal and Energy Technology Division, EGI Contracting/Engineering Co., Ltd (a privatized but formerly state-owned energy consulting firm).

Prof. Tamás Németh, Institute for Soil and Agricultural Chemistry, Hungarian Academy of Science.

Miklós Poós, Head, Department of Energy, Division of Energy Coordination, Ministry for Industry and Trade; PHARE coordinator.

Anikó Radnai, Head of Environmental Impact Assessment Section, Department for Environmental Policy, Agency for Environmental Protection, Ministry for Environment and Regional Policy.

Róbert Rakics, Head, Department of Air Pollution and Noise Control, Ministry for Environment and Regional Policy; delegate to UNECE CLRTAP since 1992.

Dr Judit Rákosi, economist and manager, Öko Rt. (Eco Inc.), environmental consulting firm, Budapest.

Dr Erzsébet Schmuck, Hungarian Association of Nature Conservationists, NGO umbrella organization head.

Miklós Szoboszlay, Head, Department of Environment, Ministry for Transport, Telecommunications, and Water.

Róbert Tóth, Department Secretary and Meteorologist, Department of Air Pollution and Noise Control, Agency for Environmental Protection, Ministry for Environment and Regional Policy.

Elek Turda, EGI Contracting/Engineering Co., Ltd (a privatized but formerly state-owned energy consulting firm).

Tibor Várkonyi, Institute for Public Health; member, Interministerial Steering Group; former director of Department of Air Quality.

Anna Váry, Director, Energy Club, an NGO for environmental protection and headquarters for Elég (campaign for energy conservation) Budapest.

János Zlinszky, staff member, The Regional Environmental Center for Central and Eastern Europe.

# Bibliography

Abel, Istvan, Csermely, Agmes, Kaderják, Peter, Pavics, Lázár and Ferto, Imre (1993), 'Environmental Implications of Economic Restructuring: The Case of Hungary', Paper prepared for the Workshop on Economic Restructuring and the Environment, Budapest, 9-10 March.

Act No. II of 1976 on the Protection of the Human Environment. For air, paragraphs 23-26.

Agh, Attila (1993), 'Europeanization through Privatization and Pluralization in Hungary', *Journal of Public Policy*, Vol. 13, No. 1, Spring, pp. 1-35.

Bándi, Gyula, ed. (1993), *Environmental Law and Management System in Hungary: Overview, Perspectives and Problems*, Prepared by the experts of EMLA: Prof. Bándi, dr Sándor Fülöp, Dr Marianna Nagy, Prof. János Szhávik, and Prof. Kees Lambers (University of Groningen), Environmental Management and Law Association: Budapest.

Bándi, Dr Gyula, Faragó, Dr Tibor, and Lakos, Alojzia H. (1994), *Nemzetközi Környezetvédelmi és Természetveédelmi Egyezmények* [*International Conventions for Environment Protection and Nature Conservation*], Budapest: Környezet védelmi és Területfejlesztési Minisztérium [Hungarian Ministry for Environment and Regional Policy].

Berman, Paul (1980), 'Thinking about Programmed and Adaptive Implementation: Matching Strategies to Situations', in Helen M. Ingram and Dean E. Mann (eds), *Why Policies Succeed or Fail*, Sage: Beverly Hills, Calif.

Bernauer, Thomas (1995), 'The Effect of International Environmental Institutions: How We Might Learn More', *International Organization*, Vol. 49, No. 2, Spring, pp. 351-377.

Bochniarz, Zbigniew, Bolan, Richard, Kerekes, Sándor, Kindler, Jozsef, Vargha, Janos, and Witzke, Harald von, with assistance from Andrzejj Kassenberg and Vaclav Mezricky (1992), *Environment and Development in Hungary: A Blueprint for Transition*, Budapest - Minneapolis.

Brown, E.D., and Churchill, R.R., eds (1985), *The UN Convention on the Law of the Sea: Impact and Implementation*, Law of the Sea Institute, University of Hawaii.

Bryner, Gary (1996), 'Inequality, Environmental Justice, and the Implementation of Global Environmental Agreements', paper presented at the annual meeting of the American Political Science Association, San Francisco, August.

Bulla, Miklós (1992), 'Natural Resources, the State of the Environment', Appendix I in Bochniarz et al., pp. 65-76.

Chayes, Abram, and Chayes, Antonia (1993), 'On Compliance', *International Organization*, Vol. 47, No. 2, Spring, pp. 175-205.

Comisso, Ellen, and Hardi, Peter (1997), 'Hungary: Political Interest, Bureucratic Will' in Harold K. Jacobson and Edith Brown Weiss (eds), *Engaging Countries*, Cambridge: MIT Press.

Durant, Robert F. (1985), *When Government Regulates Itself: EPA, TVA, and Pollution Control in the 1970s*, Knoxville: University of Tennessee Press.

Economic Commission for Europe (1995), *Strategies and Policies for Air Pollution Abatement: 1994 Major Review Prepared under the Convention on Long-range Transboundary Air Pollution*, New York and Geneva: United Nations.

Environmental Council of the Parliament, Government of Hungary (1995), *Report on the Activities of the Ministry of Environment and Regional Planning for the Period 1994-1995* (in Hungarian), Edited by the Strategic Department of the Ministry of Environment and Regional Policy, November.

Faragó, Tibor, and Lakosné Horváth, (Alojzina) (1995), *The Ratification and the Prosecution of International Agreements in Hungary* (in Hungarian), Ministry of Environment and Environmental Policy, Government of Hungary.

Fauteux, Paul (1990), 'Percentage Reductions Versus Critical Loads in the International Legal Battle Against Air Pollution: A Canadian Perspective', paper presented to a symposium on Environmental Protection and International Law, Vienna, 11-12 October.

Flaherty, Margaret Fresher, Rappaport, Ann, and Hart, Maureen, with contributions from Arpad von Lazar and Sándor Kerekes (1993), 'A n Environmental Brief: Privatization and Environment in Hungary. A Case Study', Center for Environmental Management, Tufts University, Spring.

Haas, Peter M., Keohane, Robert O., and Levy, Marc A., eds (1993), *Institutions for the Earth: Sources of Effective International Environmental Protection*, Cambridge, Massachusetts: The MIT Press.

Hanf, Kenneth I. (1994a), *The International Context of Environmental Management from the Negotiating Table to the Shop Floor*, Breukelen, The Netherlands: Nijenrode University Press.

Hanf, Kenneth (1994b), 'The Political Economy of Ecological Modernization: Creating a Regulated Market for Environmental Quality', in M. Moran and T. Prosser (eds), *Privatization and Regulatory Change in Europe*, Buckingham Philadelphia: Open University Press, pp. 126-55.

Hanf, Kenneth, Hjern, Benny, and Porter, David O. (1978), 'Local Networks of Manpower Training in the Federal Republic of Germany and Sweden', in Hanf and Fritz W. Scharpf (eds), *Interorganizational Policy Making*, Beverly Hills, Calif.: Sage.

Hanf, Kenneth, and Roijen, Marcel (1994), 'Water Management Networks in Hungary: Network Development in a Period of Transition', *Environmental Politics*, Vol. 3, No. 4, December, special issue.

Hanf, Kenneth, and Underdal, Arild (1996), 'Domesticating International Commitments: Linking National and International Decision Making', in Oran Young (ed.), *The International Political Economy and International Institutions*, Vol. II, Cheltenham: Edward Elgar, pp. 1-20.

Hanf, Kenneth, Andresen, Steiner, Boehmer-Christiansen, Sonja, Kux, Stephan, Lewanski, Rodolfo, Morata, Francesc, Skea, Jim, Sprinz, Detlef, Underdal, Arild, Vaahtoranta, Tapani, and Wettestad, Joergen (1996), 'The Domestic Basis of International Environmental Agreements: Modelling National/International Linkages', final report to the European Commission, EC Contract EVSV-CT92-0185, May.

Helm, Carsten, and Sprinz, Detlef (1995), 'Negotiating the Acid Rain Regime - A Historical Primer, the Functions of Intergovernmental Organizations, and Negotiation Theory', Unpublished manuscript, Potsdam institute for Climate Impact Research, Potsdam, Germany.

Hesse, J.J. (ed.), (1993), 'Administrative Transformation in Central and Eastern Europe', *Public Administration*, special issue Vol. 71, Nos. 1/2, Spring/Summer.

Hinrichsen, D., and Enyedi, G. (eds), (1990), *State of the Hungarian Environment*, Budapest: Statistical Publishing House.

Hull, Chris., with Hjern, Benny (1987), *Helping Small Firms Grow*, London: Croom Helm.

Hungarian Commission on Sustainable Development (1994a), *Energy Use and Carbon-Dioxide Emissions in Hungary and in the Netherlands: Estimates, Comparisons, Scenarios. Contribution to the National Energy and Environmental Planning in Relation to the Energy-Climate Issues*, Hungarian Ministry for Environment and Regional Policy and Netherlands' Ministry of Housing, Physical Planning and Environment.

Hungarian Commission on Sustainable Development (1994b), *Hungary: Towards Strategy Planning for Sustainable Development*, National information to the United Nations Commission on Sustainable Development, Budapest: Hungarian Commission on Sustainable Development.

Jacobson, Harold K., and Weiss, Edith Brown, eds (1997), *Engaging Countries*, Cambridge: MIT Press.

Kaderják, Péter (1993), 'A Study on Mobile Source Air Pollution Problems in Hungary', Working Paper 1993/2, Budapest University of Economics, Department of Business Economics.

Keohane, Robert O., Haas, Peter, and Levy, Marc (1993), 'The Effectiveness of International Environmental Institutions', in Haas, Keohane, and Levy (eds), pp. 3-24.

Kerekes, Sándor (1993), 'Economics, Technology, and Environment in Hungary', *Technology in Society*, Vol. 15, pp. 137-47.

Kimball, Leo A. (1992), *Forging International Agreement*, Washington, DC: World Resources Institute.

Kotov, Vladimir, and Nikitina, Elena (forthcoming), 'LRTAP: Implementation and Effectiveness in Russia', in David G. Victor,

Kal Raustiala, and Eugene B. Skolnikoff (eds), *Implementation and Effectiveness of International Environmental Commitments*, Cambridge: MIT Press.

Lammers, J. (1991), 'The European Approach to Acid Rain', in M.D. Barstow (ed.), *International Law and Pollution*, Philadelphia: University of Pennsylvania Press.

Levy, Marc A. (1993), 'European Acid Rain: The Power of Tote-Board Diplomacy', in Haas, Keohane, and Levy (eds), pp. 75-132.

Levy, Marc A. (1995), 'International Cooperation to Combat Acid Rain', *Green Globe Yearbook 1995*, Oxford: Oxford University Press.

Lindblom, Charles E. (1977), *Politics and Markets*, New York: Basic Books.

Lindblom, Charles E. (1982), 'The Market as Prison', *Journal of Politics*, Vol. 44, No. 2, May, pp. 324-36.

McCormick, J. (1989), *Acid Earth: The Global Threat of Acid Pollution*, 2d ed., London: Earthscan.

Matláry, Janne Haaland (1994), 'The European Union and the Visegrád Countries: The Case of Energy and Environmental Policies in Hungary', *Regional Politics & Policy*, Vol. 4, No. 1, Spring, pp. 136-52.

Mazmanian, Daniel, and Sabatier, Paul (1990), *Implementation and Public Policy: With a New Postscript*, Latham, Md: University Press of America.

Ministry for Environment and Regional Policy, Department for Air, Water and Soil Quality Protection, Government of Hungary (1991a), *Clean Air Protection Strategy and Concept*, Budapest, February.

Ministry for Environment and Regional Policy, Government of Hungary (1991b), *National Report to United Nations Conference on Environment and Development, 1992*, Budapest, December.

Ministry for Environment and Regional Policy, Government of Hungary (1994a), *The Central Environmental Protection Fund*, Budapest: Ministry.

Ministry for Environment and Regional Policy, Government of Hungary (1994b), *Hazánk Környezeti Állapotának Mutatói* [*Environmental Indicators of Hungary*], Budapest: Ministry.

Ministry for Environment and Regional Policy, Government of Hungary (1995), *National Environmental and Nature Conservation Policy Concept*, Budapest.

O'Toole, Laurence J., Jr (1994), 'Privatization in Hungary: Implementation Issues and Local Government Complications', in Hans Blommenstein and Bernard Steunenberg (eds), *Governments and Markets: Economic, Political and Legal Aspects of Emerging Markets in Central and Eastern Europe*, Leiden: Kluwer Academic, pp. 175-94.

Öko Rt (1992), *Study on the Short and Medium Term Environmental Action Plan of the Government* (in Hungarian), Budapest, October.

Organization for Economic Co-operation and Development (OECD) (1993), *OECD Economic Surveys: Hungary 1993*, Centre for Co-operation with Economies in Transition, Paris: OECD, September.

Porter, David O. (1976), 'Federalism, Revenue Sharing and Local Government', in Charles Jones and Robert Thomas (eds), *Public Policy Making in the Federal System*, Beverly Hills, Calif.: Sage.

Regional Environmental Center for Central and Eastern Europe (1994a), *National Environmental Protection Funds in Central and Eastern Europe: Case Studies of Bulgaria, the Czech Republic, Hungary, Poland and the Slovak Republic*, Budapest: Regional Environmental Center, November.

Regional Environmental Center for Central and Eastern Europe (1994b), *Strategic Environmental Issues in Central and Eastern Europe, Volume 1, Regional Report*, Budapest: Regional Environmental Center, May.

Regional Environmental Center for Central and Eastern Europe (1994c), *Strategic Environmental Issues in Central and Eastern Europe, Volume 2, Environmental Needs Assessment in Ten Countries*, Budapest: Regional Environmental Center, May.

Regional Environmental Center for Central and Eastern Europe (1994d), *Use of Economic Instruments in Environmental Policy in Central and Eastern Europe: Case Studies of Bulgaria, the Czech Republic, Hungary, Poland and the Slovak Republic*, Budapest: Regional Environmental Center, December.

Regional Environmental Center for Central and Eastern Europe (1995), *Government and Environment: A Directory of Governmental Organizations with Environmental Responsibilities for Central and*

*Eastern Europe*, Budapest: Regional Environmental Center, September.

Rondinelli, Dennis A., and Fellenz, M.R. (1993), 'Privatization and Private Enterprise Development in Hungary: An Assessment of Market-Reform Policies', *Business and the Contemporary World*, Vol. 5, No. 4, Autumn, pp. 75-88.

Sabatier, Paul A., and Jenkins-Smith, Hank C. (1993), *Policy Change and Learning: An Advocacy Coalition Approach*, Boulder, Colo: Westview Press.

Sand, Peter H. (1990), *Lessons Learned in Global Environmental Governance*, Washington, DC: World Resources Institute.

Scharpf, Fritz W., Reissert, Bernd, and Schnabel, Fritz (1976), *Politikverflechtung: Theorie und Empirie des kooperativen Föderalismus in der Bundesrepublik*, Kronberg: Scriptor.

Slocock, Brian (1996), 'The Paradoxes of Environmental Policy in Eastern Europe: The Dynamics of Policy-Making in the Czech Republic', *Environmental Politics*, Vol. 5, No. 3, Autumn, pp. 501-525.

Susskind, Lawrence E. (1994), *Environmental Diplomacy: Negotiating More Effective Global Agreements*, New York: Oxford University Press.

Szabo, Gabor (1993), 'Administrative Transition in a Post-Communist Society: The Case of Hungary', *Public Administration*, Vol. 71, Nos. 1/2, Spring/Summer, pp. 89-103.

Tang, Wenfang (1993), 'Post-Socialist Transition and Environmental Protection', *Journal of Public Policy*, Vol. 13, No. 1, pp. 89-109.

UNECE [United Nations Economic Commission for Europe] (1996), *Economic Development and the Environment*, New York and Geneva, United Nations.

Underdal, Arild (1979), 'Issues Determine Politics Determine Policies: The Case for a "Rationalistic" Approach to the Study of Foreign Policy Decision-Making', *Cooperation and Conflict*, Vol. 14, No. 1, pp. 1-9.

United Kingdom, Select Committee on The European Communities, House of Lords (1995), *Environmental Issues in Central and Eastern Europe: The Phare Programme*, with evidence, Session 1994-95, 16th Report, HL Paper 86, London: HMSO, 18 July.

Várkonyi, Tibor, and Kiss, Gyozo (1990), 'Air Quality and Pollution Control', in *Environment and Society*, pp. 49-65.

Vastag, Gyula, Rondinelli, Dennis A., and Kerekes, Sándor (1994), 'How Corporate Executives Perceive Environmental Issues: Comparing Hungarian and Global Companies', Publication Series, International Private Enterprise Development Research Center, Kenan Institute of Private Enterprise, University of North Carolina at Chapel Hill.

Verebelyi, Imre (1993), 'Options for Administrative Reform in Hungary', *Public Administration*, Vol. 71, Nos. 1/2, Spring/Summer, pp. 105-120.

Victor, David G., Raustiala, Kal, and Skolnikoff, Eugene (eds), (forthcoming), *Implementation and Effectiveness of International Environmental Commitments*, Cambridge: MIT Press.

Wilson, James Q. and Rachal, Patricia (1977), 'Can the Government Regulate Itself?', *Public Interest*, Vol. 46, Winter, pp. 3-14.

Young, Oran R. (1989), *International Cooperation: Building Regimes for Natural Resources and the Environment*, Ithaca, NY: Cornell University Press.

Young, Oran R. and von Moltke, K. (1994), 'The Consequences of International Environmental Regimes: Report from the Barcelona Workshop', *International Environmental Affairs*, Vol. 6, No. 4, pp. 348-370.